THE
ECONOMISTS

THE
ECONOMISTS

LEONARD SILK

Basic Books, Inc., Publishers

NEW YORK

Excerpts from *Creative Tension: The Life and Thought of Kenneth Boulding* reprinted by permission of the University of Michigan Press, copyright © 1974 by the University of Michigan.

Library of Congress Cataloging in Publication Data

Silk, Leonard Solomon, 1918-
  The economists.

  Includes bibliographical references and index.
  1. Economists—United States—Biography. I. Title.
HB199.A3S54   330'.092'2 [B]   76-27741
ISBN: 0-465-01810-6

TO MY TEACHER

Joseph J. Spengler

CONTENTS

PREFACE

JOHN MAYNARD KEYNES, writing in the midst of the Great Depression of the 1930s, prophesied that the ideas he was setting forth in his magnum opus, *The General Theory*, would, if correct, have "potency over a period of time." People were then desperately waiting for a more fundamental diagnosis of the economic ills of the time, and were eager to try it out. Apart from that contemporary mood, said Keynes,

> . . . the ideas of economists and political philosophers, both when they are right and when they are wrong, are more power-ful than is commonly understood. Indeed the word is ruled by little else . . . I am sure that the power of vested interests is vastly exaggerated compared with the gradual encroachment of ideas. Not, indeed, immediately, but after a certain interval; for in the field of economic and political philosophy there are not many who are influenced by new theories after they are twenty-five or thirty years of age, so that the ideas which civil servants and politicians and even agitators apply to current events are not likely to be the newest. But soon or late, it is ideas, not vested interests, which are dangerous for good or evil.

In this book I have sought to explore the ideas of five lead-ing economists of the generation that followed Keynes. All of them have produced ideas which have already had a consid-erable impact upon contemporary society, ideas which I be-lieve will continue to influence economic and political thought and action for many years to come.

So great has been the growth of public interest in the ideas of economists since the Depression that there is no longer the same degree of nebulousness surrounding the communication of thoughts and policy proposals from economists to politicians and the general public. In his day Keynes said that "practical men, who believe themselves to be quite exempt from any intellectual influences, are usually the slaves of some defunct economist. Madmen in authority, who hear voices in the air, are distilling their frenzy from some academic scribbler of a few years back." But nowadays practical men and politicians, mad or sane, whether in authority or only seeking to attain it, are far more likely to talk directly with economists, to distill their frenzy and their programs from the thoughts expressed by learned men sitting face to face across a table. Sometimes the ideas of the economists are set forth in memoranda that then go into the political meatgrinder together with other political ideas, polls, and pressures.

All five economists portrayed in this book have been politically *engagé*, but none has been a mere handyman for politicians. All are eminent scholars whose ideas are likely to go on reverberating beyond the confines of the economics profession and their current political context. Each has been a president of the American Economic Association, and two— Paul Samuelson and Wassily Leontief—have won the Nobel Memorial Prize in Economic Science. I have chosen these five among a larger number of distinguished economists because they seemed to me best to represent the broad mainstream of the American economics profession and because they have addressed themselves not only to technical problems but to the deepest issues troubling contemporary society. Those issues include the traditional ones of economic growth and stability—inflation and unemployment, sometimes combined in the difficult mixture now called "stagflation"—as well as the heightened tensions and threats to peace that emerge from inequality in the distribution of wealth and income among individuals and nations; the threats to the natural and social environment of an ever-expanding industrial civilization;

the perplexing issues of social welfare—what it is and how to increase it; the competing or complementary roles of market forces and government planning or control. Underlying all these issues and others lie the moral and ethical concerns from which economics has never succeeded in divorcing itself.

It has been my hope that by exploring the ideas, the careers, and, to a degree, the personalities of these five econmists, all of whom I have known for many years, I might give a picture of the present state of economics. Within the American economics establishment, I consider Paul Samuelson to be the vital center, with intellectual and moral views which themselves extend over a wide spectrum and are difficult to categorize. Nevertheless, clearly to his right is Milton Friedman, to his left John Kenneth Galbraith and Wassily Leontief. Kenneth Boulding does not really fit conveniently on any left-right axis; his ideas are deeply religious, as well as economic and scientific. He is reaching for a social science that may one day make economics itself obsolete.

Economics, as the late Professor Jacob Viner of Princeton first remarked, is what economists do. This is not merely a truism or an evasion of the tricky task of defining this subtle, arcane but not necessarily dismal science. For economists, like other creative artists or scientists, are not bounded in their work by the formal ideas of one discipline. Their lives, their philosophies, and their wide-ranging perceptions explain much that goes beyond the explicit logic of their professional work. I feel that this is particularly true of the five complex and fascinating personalities here under consideration. In a final chapter, after the five biographical essays, I offer some comments on the present state—and deficiencies—of contemporary economics, as exemplified by the contributions of these economists.

In writing this book, I have had the enormous help of my oldest son, Mark, a graduate student and tutor of History at Harvard. He did some of the most thoughtful and imaginative library work and legwork that any author could ask. The chapters on Friedman, Galbraith, and Boulding particularly benefited from his endeavors.

I wish to thank Emma Rothschild and Robert B. Silvers, editor of *The New York Review of Books*, for making their taped interviews with Wassily Leontief available to me.

I am also grateful to Amelia North, who did sharp detective work for the Leontief chapter, and to Jean Kidd, who ably handled the manuscript.

This book would not have been written without the original conception and urging—and the later encouragement and suggestions—of Martin Kessler, my editor. Every author should be so lucky!

Finally I wish to thank Professors Samuelson, Friedman, Galbraith, Leontief, and Boulding for their cooperation and patience.

Montclair, New Jersey
May 30, 1976

Paul Anthony Samuelson

Enfant Terrible Emeritus

" 'Bourgeois,' I observed, 'is an epithet which the riff-raff apply to what is respectable, and the aristocracy to what is decent.' "

—*Anthony Hope*

I

IN ECONOMICS, a field where white hair and an endowed chair are often taken as vouchers for wisdom, Paul Samuelson was a wonder child. His natural gift for economics was astonishing: most of his *Foundations of Economic Analysis* was written in 1937 when he was a 22-year-old graduate student at Harvard. In that work he sought to clean out the Augean stable of economics, which he found full of contradictions, overlaps, fallacies, and other detritus. He told his fellow economists that they had been practicing "mental gymnastics of a peculiarly depraved type" and that they were like "highly trained athletes who never run a race."

Though so fast off the mark himself, Samuelson proved to have the heart and drive of the long-distance runner. He has done it all, exploring microeconomics, macroeconomics, welfare economics, international monetary theory and, as he says, "turnpike theorems and osculating envelopes; nonsubstitutability relations in Minkowski-Ricardo-Leontief-Metzler matrices of Mosak-Hicks type; balanced-budget multipliers under conditions of balanced uncertainty in locally impacted topological spaces and molar equivalences." But he has also written the best-selling economics textbook of all time, written for the newspapers, been a magazine columnist, advised presidents, and won the Nobel Memorial Prize in Economic Science—the first American to get it.

Once regarded as a brash, arrogant opponent by the pillars of the economics establishment, he has lived to embody that establishment in his own person. The attack on— or defense of—contemporary economics must begin with

[3]

Paul Samuelson, whom time and his own talent and industry have endowed with fame, wealth, and respectability. He has become the leading practitioner of what may be called bourgeois economics.

Samuelson regards economics not as a bourgeois science but as a science essentially of the bourgeoisie. "Economics," he says, "seems to decay under a nonbourgeois society." He notes that when two men whom he admired as creative and productive scholars, the economists Oskar Lange and Michal Kalecki, returned to Communist Poland and Czechoslovakia after World War II, neither did scientific work comparable to what each had done earlier as refugees in the capitalist West. And Samuelson is not impressed with the Soviet economists he meets, "unless they are simply applied mathematicians, in which case they can be extremely good, as in probability theory."

"All that," says Samuelson, with his unique blend of pride and wry self-deprecation, "is just a way of building up to the fact that my own family background is middle class. Usually when you have a scholar of some distinction, he turns out to be descended from a long line of rabbis. I don't know what happened further back, particularly since my mother always had some delusions of grandeur about her lineage. But as far as I know, this had no justification at all, except that her grandfather, who came to this country before the Civil War at the time of the Gold Rush, got a nest egg that put her family a little bit higher in the social scale. By miscalculation, he went back to Europe—to that part of Poland that abuts on East Germany."

Samuelson's own father was a druggist who had a store in Gary, Indiana, where Paul was born in 1915. "He did very well in the new frontier town," says Samuelson, "the company town created by U.S. Steel, where coal meets ore at the tip of Lake Michigan. Then along came World War I, and I learned about the Keynesian multiplicand in my mother's milk, so to speak, because the steel mills flourished and that meant my father's drugstore in Gary also flourished."

Samuelson talks of the "bourgeois genes" he inherited from both sides of his family. He tells of visiting Knott's Berry Farm in California with his "ancient mother" and his wife and six children. "It's a place where people from Los Angeles go to get a chicken dinner and ride on an old engine, and there's a mining village—the nearest thing to early times that you have in California. As we were looking at this old mining town, I said, 'Well, Mother, you always talked about Grandfather who was a Forty-niner. What in the world did he do out here?' And she said, 'What do you think he did? He sold trinkets to the Indians.'"

Samuelson thinks it is not coincidental that a considerable number of good economists are Jewish. But he concedes that the explanation may be more Marxian than Darwinian. "Following Marx's ideas about the material conditions of production guiding the superstructure of ideology and analysis and choice of subject," he says, "Jews have been entrepreneurial for various reasons, not all of their own volition, that may have enabled them to become good economists and predisposed them to understand money." (Milton Friedman, Samuelson's distinguished contemporary and intellectual rival, thinks the market "has been good for Jews" in economics and other professions, in breaking down the barriers of caste and custom.)

But Samuelson feels that there is another, disparate element in the Jewish tradition that may help explain the affinity between the Jews and economics—what he calls the "troublemaking aspect." "The Jew, as a stranger," he says, "is somewhat alienated from the culture he's in, and therefore he's more critical." Thorstein Veblen, the heterodox Norwegian-American economist, expressed the same thought half a century ago:

The Jew who searches after learning must go beyond the pale of his own people. But although he finds his own heritage untenable, he does not therefore take over and assimilate the traditions of usage and outlook which the new role has to offer. The idols of his own tribe have crumbled in decay and no longer cumber the ground, but his release from them does not induce him to set up a new line of idols borrowed from an alien tribe.

Intellectually, he is likely to become an alien. Spiritually, he is more likely to remain what he was—for the heartstrings of affection are tied early and they are not readily retied in after life. Nor does the hostile reception he finds waiting for him in the community of the safe and sane make possible his personal incorporation in that community, whatever may befall the intellectual assets he brings. Their people need not become his people, nor their gods his gods. Indeed, provocation is always present to turn back from following after them.[1]

Samuelson thinks the Jewish tradition may involve "a union of opposites." At times it may lead to a success-oriented opportunism. He says, "You have to go along to get along." But he adds that this union of opposites also leads to "a certain amount of what Uncle Remus would recognize as Aesopian duplicity." "You lie low," he says, "if it is an expedient thing to lie low, and then you come out—like a good southerner who knows how to fight for the good cause, let's say, in race relations. I don't mean 'the good ol' boy,' I mean the opposite—the good southerner who knows just what the traffic will bear. He knows when he can come on hard, and when he had better beat an opportunistic retreat and lie low, just waiting for a more convenient battlefield for his next move."

Samuelson sees this union of opposites both in his father and in himself. "My father would have voted for Norman Thomas," he says. "I used to read on our family bookshelves books which my father, I guess when he was a bachelor, had acquired, many of them in used bookstores in Chicago, such as debates between Clarence Darrow and other people on pacifism and religion and so forth. I was reasonably inclined, I guess, to be New Dealish before there was a New Deal."

Samuelson's family moved from Gary to Chicago in 1923. He attended public schools and Hyde Park High School, which is now almost completely black, he says. "It's not very far from the University of Chicago and had a long tradition of sending scholars to the University of Chicago."

"I had a famous high school mathematics teacher who has trained quite a number of eminent scientists—Malcolm Smiley, the mathematician; Norman Davidson, the chemist; myself; Roy Radnor, the economist, I learned years later, was a graduate of Hyde Park High School; my brother Bob, who is an economist. This mathematics teacher, Beulah Shoe-smith, was a real terror of the old school. It turned out that she had left $2 million to the University of Chicago when she died, showing the power of thrift, since she very rarely changed her dress. I asked my brother once how she did it. He said he didn't know but he would find out and inform me. And the best he could learn was that she had a good broker. And my reaction was we've all been looking for that fellow for many years.

"In those simple days of the Depression, you went to college near your home, or to your father's college. My father had gone to pharmacy college, and his last wish was to have any of his three boys become a pharmacist. I think he saw the chain store trend handwriting on the wall. So I went to the University of Chicago.

"As I think back, it's also a reflection of how things have changed. It never would have occurred to me—as a pretty good student, although I was what came to be called an underachiever in high school, when you were ashamed to be a grind—to aim for Harvard. I don't think I knew anybody who had gone to Harvard. My wife, who comes from a small town in Wisconsin and who went to Radcliffe, never heard of Radcliffe until within half a year before she went there. Anybody who went East to school or went to a private school was usually part of a broken family, in which case the poor boys had to go to military school. Or there was, of course, a very small upper crust of people whose families had always gone to Groton and so forth, but they were practically almost unobserved by most of the population."

Thus Samuelson, a New Dealer before the New Deal, by happenstance enrolled at the University of Chicago, whose economics department was the foremost center of the laissez-faire philosophy in the United States. This set up a new tension within him—between individualism and the need for social action. His life's work can be seen as an effort to forge a new "union of opposites."

II

"I WAS BORN as an economist," says Samuelson, "on January 2, 1932, which means that chronologically I was about sixteen and a half years of age."

That was the morning when, as an entering freshman but still not out of high school, he went to an eight o'clock class to hear Louis Wirth, the eminent sociologist, talk on Malthus. It was the first lecture of his first course at the University of Chicago. "I found Malthus very sexy," he remembers.

The Malthusian theory about population growth as the cause of poverty, with the 1,2,4 . . . growth of population outrunning the 1,2,3 . . . growth of the means of subsistence, seemed so simple that he thought there must be something else to it and that later, when he grew up, he would learn what it was. "But I discovered that there wasn't anything more to it than the simple syllogism that I was imbibing that morning."

Paul Anthony Samuelson

Coming into the university one quarter late with the entering class of 1935, because he had not yet graduated midterm from Hyde Park High School, Samuelson was put into a "remedial" course in elementary economics that was being taught by Aaron Director, a Chicago libertarian so orthodox that he later came to call Milton Friedman his "radical brother-in-law." Samuelson took to Director's economics course, he says, "as a cat takes to catnip, or as my first daughter took to ice cream at the age of nine months when ice cream was put on her tongue. She went wild—there were things she had never heard of! I took to economics that way."

Samuelson liked Director. "He had a very cold, sardonic sense of humor. He was an iconoclast. This was at the bottom of the Great Depression, and to have a member of the Chicago School casting scorn upon so many of the panaceas being offered at that time was titillating."

Samuelson came alive intellectually at Chicago. Because he had arrived a quarter late, he thought he had to work extra hard just to get by. He says his situation was like that of the sociologist Robert K. Merton, who arrived at Harvard Graduate School in the 1930s from Temple University and, not knowing what was expected of him, did about five years' work in two. "I got trapped into achievement in a similar way," says Samuelson. "I liked all my subjects—economics, the physical sciences, the biological sciences, sociology, anthropology. At one moment I toyed with the notion, probably because I enjoyed Louis Wirth's lectures, of majoring in sociology." He also admired, and got to know outside of class, the mathematician Gilbert Bliss, the biologist Anton Carlson, the anthropologists Fay-Cooper Cole and Robert Redfield, the paleontologist Alfred Romer, the political scientists Frederick Schuman and Harold Lasswell, and the American historian W. T. Hutchinson, whom he still considers the best lecturer he has ever heard.

He took Jacob Viner's course in graduate economic theory, famous both for its profundity and for Viner's manhandling of students. Samuelson has said, "I, nineteen-year-old innocent, walked unscathed through the inferno and naïvely pointed out errors in his blackboard diagraming. These acts of Christian kindness endeared me to the boys in the back room of the graduate school: George Stigler, Allan Wallis, Albert Gaylor Hart, Milton Friedman, and the rest of the [Frank] Knight 'Swiss guards.' As I performed various make-work tasks for the department—dusting off the pictures of Böhm-Bawerk, Menger, and Mill in the departmental storage room which Stigler and Wallis had squatted in—we would gossip for hours over the inadequacies of our betters and the follies of princes who try to set right the evils of the marketplace."[2]

Samuelson left Chicago for his graduate work because he won a Social Science Research Council fellowship, awarded experimentally in 1935 to the eight most promising economics graduates. The fellowships, subsidizing their whole graduate training, had only one stipulation: that the fellows go elsewhere than their undergraduate colleges for their graduate studies. Samuelson chose Harvard.

III

HE CHOSE HARVARD rather than Columbia (in those days considered one of the top three schools in economics) for nonrational reasons—"in search of green ivy." He says he almost started back home after his first glimpse of Harvard Yard, approached "by bad luck from Central Square."

He also was off to a bad start with the chairman of Harvard's economics department, Harold Hitchings Burbank. The new graduate student from Chicago told the Harvard economics department chairman that he would not take E. F. Gay's famous course in economic history because he intended to "skim the cream" of Harvard, since he wasn't sure he would choose to stay more than a year. Professor Burbank also asked Samuelson why he had not made advance application for graduate study at Harvard, and Samuelson assured him that a Social Science Research Council predoctoral fellow could get in anywhere.

"It was not love at first sight," says Samuelson. "But it really would not have mattered, since Burbank stood for everything in scholarly life for which I had utter contempt and abhorrence."

Nevertheless, Samuelson flourished at Harvard. He had great teachers—Alvin Hansen, Joseph Schumpeter, Wassily Leontief, and E. B. Wilson, a mathematical statistician and physicist whom he revered. He also had a sprightly crew of young instructors and fellow graduate students as friends— John Kenneth Galbraith, Alan and Paul Sweezy, Robert Aaron Gordon, Abram Bergson, Richard Musgrave, Shigeto Tsuru, Lloyd Metzler, Robert Triffin, James Tobin, Walter and William Salant, Richard Goodwin, Henry Wallich, Sidney Alexander, and others. No one can quite explain why any single institution becomes hotter than the others, but when it starts to happen, there is a clustering of star-quality graduate students. Harvard was getting them in abundance in the late 1930s. As Samuelson was later to say, "Harvard made us. But we made Harvard."

As a graduate student at Harvard, Samuelson led the drive to clarify and systematize economic theory by the use of rigorous mathematical methods. Before his twenty-third birthday he had written a stream of brilliant articles: "A Note on Measurement of Utility,"[3] "Some Aspects of the Pure Theory of Capital,"[4] "A Note on the Pure Theory of Consumer's Behavior,"[5] "Welfare Economics and International Trade,"[6] "The Empirical Implications of Utility Analysis."[7] The stream continued, year after year, deepening as it flowed.

From the beginning, Samuelson was recognized as a

prodigy, an enfant terrible (when he grew older Professor Peter Kenen of Princeton called him an "enfant terrible, emeritus"). An article Samuelson wrote as a "finger exercise" for Professor Alvin Hansen (published in May 1939 as "Interactions between the Multiplier Analysis and the Principle of Acceleration")[8] quickly brought him worldwide fame, for he had succeeded in transforming Keynes's static analysis of the forces that depressed an economy and produced high unemployment into a dynamic description of the factors causing capitalist economies to swing up as well as down. In spelling out the interactions of consumption, investment, and national income, Samuelson demonstrated, in a few elegant pages, that the economy would not just drop until it reached a new equilibrium, but that its fall would slacken and then reverse itself. Similarly, his analysis showed that an economic expansion would slacken and turn down. Using the basic Keynesian apparatus, Samuelson showed that capitalist economic systems were inherently oscillatory but not wildly unstable. The economy as a whole was not based on some maximizing principle but was cyclically dynamic.

Samuelson remembers his days as a junior fellow at Harvard before the war as pure heaven, "with reams of blank paper, sharpened pencils, no duties, adequate pay, and best of all, most of what needed doing in economic theory not yet done." Besides all this, he had married Marion Crawford, a fellow graduate student in economics. He speaks of her as virtually the coauthor of his doctoral dissertation, *Foundations of Economic Analysis*,[9] which he still loves and has never changed as it has gone through successive printings. "A scholar's books are like his children," he has said. "They come to lead a life of their own, but he tends to remember how they were in the beginning."

He thinks of those glorious days in prewar Cambridge and says, with Talleyrand, that "only those who have known the ancien regime can appreciate its sweetness." He recalls that certain of his *Foundation*'s theorems were developed as he was walking along the Charles, some "while sitting in the front seat of a car being driven westward at 70 miles an hour by my wife." He says that he often dreamed of theorems in his sleep—and some of them were true. "In those days thinking about economics filled my every hour." He not only thought constantly but read furiously; he learned the history of economic literature as few, if any, know it today. He was not afraid of discovering that he was not the first to have a particular idea; he was not interested in the kind of "originality" that comes from not knowing the literature. On the contrary, he was proud to do justice to the literature.

But his originality shone through. He was awarded the David Wells Prize for his dissertation, so boldly titled; its publication as a book was held up by World War II, and by the time it came out in 1947, Samuelson had come to fear that his preoccupation with pure economics might be "somewhat decadent." But the book was a great and instantaneous success. This gave him a great deal of pleasure, especially when he thought of the hated "Burbie," Harold Hitchings Burbank, chairman of the Harvard economics department, who remained his foe—and not his only one at Harvard, especially among the senior faculty. Twenty years after *Foundations of Economic Analysis* was published, Samuelson wrote in a foreword to the Japanese translation: "It is sweet revenge upon a departed head of the Harvard economics department, who had to be forced into printing a thousand copies and who had the type immediately destroyed despite the expense of mathematical typesetting, that the book still continues to sell as many in each year as he had wanted to print in all."[10]

IV

IN 1940, Harvard offered Samuelson an instructorship in the economics department and a tutorship in the Division of History, Government, and Economics, which he gladly accepted. But a month later, the Massachusetts Institute of Technology offered him a professorship—and Harvard made no effort to keep him. The popular explanation at the time Harvard let Samuelson go was anti-Semitism. Harvard did try, at least twice in the postwar period, to get him back, but Samuelson declined, maintaining that after "a cost-benefit analysis" he had simply decided to stay at MIT.

He maintains that anti-Semitism is too simple an explanation for why Harvard let him go. Nevertheless, like Uncle Remus or his "good southerner," when the time was right, Samuelson permitted himself to express his disgust with the anti-Semitism that characterized not just Harvard but other American colleges and universities in the not always sweet days of the ancien regime:

> Lest I be misunderstood, let me state that before World War II American university life was antisemitic in a way that would hardly seem possible to the present generation. And Harvard, along with Yale and Princeton, was a flagrant case of this. So if anyone wants to understand why Jews, in relation to their scholarly abilities, were underrepresented on the Harvard faculty in those days, he can legitimately invoke the factor of antisemitism. (In many of the humanities faculties, bigots genuinely believed that Jews were no good; in science and mathematics, the belief was that they were too good; one could, so to speak, have one's cake and eat it too by believing—as did the eminent mathematician George D. Birkhoff, and to a degree, the eminent economist Joseph Schumpeter—that Jews were "early bloomers" who would unfairly receive more rewards than they deserved in free competition. Again, lest I be

misunderstood, let me hasten to add the usual qualification that these two men were among my best friends and, I believe, both had a genuine high regard for my abilities and promise.) To illustrate, though, the failure of any one factor to account for the richness of reality, the following questions can be asked of those who know economists of that period well. (1) Why was Lloyd Metzler not given a tenure post at Harvard, since he suffered only from the disability of being from Kansas? (2) If you contemplate the academic careers in America of three men—Oskar Lange, Jacob Marshak, and Abba Lerner—how can you cover the facts with the simple theory of antisemitism?[11]

Harvard may have had other reasons for not trying to keep him. He had certainly strengthened the case against himself by his inability to suffer fools, especially among his academic superiors. He recognizes his own tendency to test his mettle against the best minds (as he did, as an undergraduate, against Jacob Viner), and he knows that he has not won every battle. In his Nobel Prize address (though warned by the Swedish chairman, Professor Erik Lundberg, that he had to be serious), Samuelson told of his encounter with "Goliath"—the great mathematician John von Neumann. "Sometime around 1945," Samuelson told the glittering audience in Stockholm, "von Neumann gave a lecture at Harvard on his model of general equilibrium. He asserted that it involved new kinds of mathematics which had no relation to the conventional mathematics of physics and maximization. I piped up from the back of the room that I thought it was not all that different from the concept we have in economics of the opportunity-cost-frontier, in which for specified amounts of all inputs and all but one output society seeks the maximum of the remaining output. Von Neumann replied at that lightning speed which was characteristic of him: 'Would you bet a cigar on that?' I am ashamed to report that for once little David retired from the field with his tail between his legs.

And yet some day when I pass through Saint Peter's Gates I do think I have half a cigar still coming to me—only half because von Neumann also had a valid point."[12]

As difficult (or arrogant) as Samuelson had been toward his masters at Chicago and Harvard, he was kind and generous to his students at MIT. He blossomed as a teacher and helped to make the Institute's economics department one of the very best in the country. Although he spent part of the war years working in MIT's Radiation Laboratory, developing computer systems for tracking aircraft and missiles, and later did consulting work for the War Production Board and the Treasury, Samuelson kept up a steady output of scientific work on a vast range of topics in both economics and pure mathematics.

At the end of the war, Samuelson took on the chore at MIT of teaching the basic principles course in economics; out of this came his textbook, *Economics*.[13] It has become the best-selling economics text of all time. When it was first published, Professor George Stigler of the University of Chicago said, "Professor Samuelson, having achieved fame, now seeks fortune." The book has sold over 3 million copies and made millions of dollars for its author and his family.

V

WHEN SAMUELSON'S *Economics* first appeared after the war, it sent shock waves through the economics world, for it was the first textbook to teach beginning students Keynesian principles. Samuelson's scientific prestige helped to sell the text, as Keynes's own prestige had won a hearing for ideas in *The General Theory of Employment, Interest, and Money* that had been regarded (and dismissed) for generations as false heresies. Young professors, children of the Keynesian revolution, delighted in Samuelson's wit and originality in presenting the then radical idea that unemployment could be cured by heavy government spending or tax cutting that produced unbalanced budgets—and the symmetrical argument that inflation could be stopped by reducing government spending, raising tax rates, and swinging the budget into balance or surplus.

Economists attending international conferences in the postwar years often found that the one thing they had in common was that they had learned their economics from Samuelson. His book has been published in French, German, Italian, Spanish, Korean, Japanese, Chinese, Hungarian, Polish, and other languages; it has often been pirated—both outright and, less obviously, in the textbooks of others. Some economists feel that Samuelson's textbook is no mere by-product of his serious work but may represent his greatest single contribution, in giving the world a common economic language.

Many conservatives hold Samuelson's "popularizing of Keynes" responsible for causing much of the continuous postwar inflation. They claim that his work fostered an enlarged role for the state by legitimating budget deficits as a cure for unemployment. One department store owner went around the United States lecturing to Rotary Club audiences about Samuelson's dangerous doctrines. Some businessmen campaigned against his textbook's use in colleges. The right-wing editor and columnist William F. Buckley, Jr., attacked Samuelson's influence in his first successful book, *God and Man at Yale*, "under the first heading of which," says Samuelson, "the New Haven chaplain came in for a lot of criticism and under the second heading of which I received my due."

Samuelson spent countless hours responding to these conservative attacks, defending his book before boards of regents and at universities, preparing mimeographed sheets citing his critics' misquotations, all to prevent the book from being dropped as "radical." Samuelson says his last wish was to have "an intransigent formulation that would be read by no one," but he would not give up the substance of the economic ideas he was trying to impart. His basic defense was to aim for empirical accuracy and logical cogency. But he concedes that such defensive writing weakens the élan of a book. His ten editions go through a cycle in their degrees of élan; they are strongest in the earliest editions, grow more cautious in the middle editions, and more daring in the end.

Samuelson has sought to avoid being pushed into a doctrinaire ideological position. When the University of Chicago sought to attract him away from MIT in 1948, the chairman of Chicago's economics department, Theodore Schultz, told him, "We'll have two leading minds of different philosophical bent—you and Milton Friedman—and that will be fruitful." Samuelson thought that one over and said no. He thought the change and the debate would be wasteful of his time. He told Professor Schultz that "it would polarize me, it would radicalize me in a way I didn't want to be radicalized." He was afraid that, "just in reaction against something which, to my mind, was too far right, I would find myself adopting positions too far left, too far from the balanced truth, which I occupy right at Aristotle's golden mean."

As the years have gone by, Samuelson has felt more pressure from the left than from the right. He contends that most of the younger people who would call themselves "new leftists" or "radical economists" are not really nonconformists "in the sense that they had a reactionary father and are slaying that father every day when they punch a dean." On the contrary, he believes that they are almost always the children of successful professional people with rather liberal viewpoints. He suggests that the young people of the left have been conditioned by their own past environment and are carrying on their parents' struggle in the new environment—

"Clausewitz's same old war," he says, "being carried on by other means."

Samuelson feels that his own family background gave him a good start toward being heterodox in his views, but this was countered not so much by a feeling of complacency or identification with the current order as by something typical of his whole scientific style. "I rebel," he says, "against any kind of romantic and wishful thinking." He liked the same quality in John F. Kennedy, who had sought Samuelson out in the late 1950s.

"When I was an adviser to Senator Kennedy, and candidate Kennedy, and President-elect Kennedy, and in a more peripheral way to President Kennedy," recalls Samuelson, "one of my functions was to harp on the need for an expansionary fiscal policy. It was a bad moment when Kennedy told his economic advisers that they couldn't have a tax cut because he had promised Senator Kerr and Speaker Rayburn that he would not call for a tax cut. Well, the problem, as an adviser, is to go as far as the traffic will bear. The traffic would bear only a limited load in the case of John F. Kennedy.

"In the first place, you couldn't just repeat an argument because after you had done that a couple of times, he would say, 'Yes, I understand that, so don't use that argument again.' But one of the things that we would urge on him was, 'Well, fight the good cause, and if you lose, you've fought the good cause.' His answer on one occasion was, 'Well, no. That's vanity. If I just do it to fight the good cause and be able to say I was on the right side, and lose, and know I'm going to lose, that just jeopardizes something else which I could get.'

"Well," Samuelson concludes, "that is the kind of argument that had a very resonant response in me, and it really shut me up."

At bottom, he stresses, it is a matter of style. "I have very few words that I have written—and I have written too many —that I have to eat. Some people—again it's a matter of style—just have different modes of speech. Milton Friedman, for instance, will say, 'I cannot conceive that the 1970 recession will not be worse than the recession of 1960–61 and as bad as that of 1957–58.' That simply means, as I read Milton—having been his friend for forty years, I guess—that he rather thinks that's going to happen. But his way of putting it is that it is *certain* to happen, that every sensible man must agree it is going to happen. Now my lips can't be brought to say that. It's part of my code to try to indicate the spectrum of uncertainty in things.

"That is very bad for a Utopian reformer, and it is that element that has been dominant in my own thinking. People don't change much in their fundamental set after the age of thirty. There are exceptions. Leontief, an old teacher of mine whom I have known and liked over the years, is an exception in that he has become more radical. He was almost completely apolitical when I first knew him and for years afterward. But he is an exception."

Samuelson feels that his own mind-set has changed but little over the years. He thinks he may have been "somewhat brainwashed" at the University of Chicago: "I was brought up by Jesuits, so I know what a Jesuitical training is. But most of that has worn off."

Though he has remained resistant to the Chicago School's almost complete trust in the marketplace and its laissez-faire philosophy, Samuelson has also grown wary of putting too much trust in government. One of the main changes in his political philosophy was produced by Senator Joseph Mc-Carthy of Wisconsin and his "witch-hunting" techniques. "McCarthy," he says, "had a big effect on me. McCarthy, I think, was a near brush with fascism in this country. I don't know why, but Richard Nixon seems to me to have been a far brush. It might have gone further but it didn't.

Paul Anthony Samuelson

"It was an honor for me to be on Nixon's 'enemies list'—
even as a journalist, as I was put in [that is, as a columnist
for *Newsweek*] rather than an academic. . . . But Joseph
McCarthy actually drove me a little to the right, as people
vulgarly measure these things. It wasn't anything he did to
me, but what I saw as the horrors of a one-company town,
when people who lost their job . . . couldn't get another job,
because they were blacklisted. Centralization in government
narrows the pluralistic opportunities."

Samuelson says that with time—and "what's called judgment in some circles and senility in others"—he has realized
the gap between high pretensions and probable performance,
whether public or private. His concern over the dangers of all-
powerful government do not make him an enthusiastic supporter of the social ethics of the private economy or even of
the private university when left to the powers of a self-serving
elite. He notes that Hollywood film producers can have a
blacklist as well as the State Department. He remembers that
Jewish economists, excluded from private colleges and universities in the prewar years, wound up working for the
Commerce Department, the Labor Department, and other
government agencies.

"You know," he says, "the one person who refused to sign
the Teachers' Oath in Massachusetts before World War II—
when the Kirtley Mathers and other leaders backed down in
order to hold their jobs, or, as they said, to spare Harvard the
penalty of their position—was a Quaker, a man named Win-
slow, who was chairman of the economics department at
Tufts College. And was he rewarded? Well, perhaps in
Heaven. But he didn't get another academic job. The frontier
for him was the Tariff Commission, where he went and died."

"But," says Samuelson, weighing the issue yet again, "that
also tells us something about government—and what's
changed."

VI

THE ABILITY to sustain tension among alternatives and the scrupulous search for resolutions on the basis of logic and empirical fact are the essence of Samuelson's economics, as well as of his social and political philosophy.

He is always willing to reconsider arguments no matter how unsympathetically he may once have regarded them. For example, in 1945 at Harvard he heard Friedrich von Hayek, the distinguished economist and philosopher, whose views are far more conservative than Samuelson's, deliver a paper on the uses of information in society. Hayek maintained that one of the tremendous advantages of a free market over a planned economy is that the voluntary cooperation of individuals utilizes far more information and knowledge than any single individual possesses or than the government can possess. Individuals, he argued, adjust their activities on the basis of how others' actions affect their own situations; and the interactions and voluntary cooperation among individuals give rise to better and wiser social institutions than the state can create. Hayek argued, in effect, that human society is simply too complex for state planners. To understand and find the best solutions to social questions, the government would have to know what all the separate people know, and that is beyond the reach of any one person or any one authority. Thus Hayek maintained that, in economics, the power of anyone to analyze, predict, and control social events is permanently limited, and attempts to disregard such limits must lead to serious error.

Originally, Samuelson thought the paper "pretty poor, pretty small beer." But by 1975, Samuelson had decided that it was Hayek's "best contribution." Samuelson said he had discovered that Hayek had "increased in wisdom over the years, the way Mark Twain's father got smart in the years between Mark's fourteenth and twenty-first birthdays."

However, despite his increased respect for Hayek's work on information, Samuelson still holds to the conclusion he reached in a 1956 paper, "Intertemporal Price Equilibrium: A Prologue to the Theory of Speculation,"[14] that additional bits of information in a market were worth *something*, but not nearly as much as the speculator was able to extract from those bits of information in relation to their social value.

Admirers of capitalism have justified speculative profits on three bases: (1) a person "deserves" the profits from his successful speculations as a reward for his knowledge and daring; (2) the greater the number of voluntary speculators the better, since society will then have fuller access to the information and judgments it can benefit from; and (3) with free entry into the business of competitive speculation, the profits of all speculators will diminish, and sum to zero or to the "right amount" proportional to the value of all speculators' information and actions.

None of these statements is valid, said Samuelson in 1956. All that economic theory really said was that the equilibrium pattern resulting from the actions of all speculators would be better than any other outcome. But this did not enable one to identify the contribution of any speculator's single act or collection of acts. Samuelson, who has been not just a fascinated observer but an active participant in speculative securities markets, like Keynes, asked: "Suppose my reactions are not better than those of other speculators but rather just one second quicker?" This, he said, might be because of his "flying pigeons" or the quickness of his neurones. In a world of uncertainty, he would note the consequences of each changing event one second faster than anyone else, and make his fortune—not once, but every day that important events happen:

Would anyone be foolish enough to argue that in my absence the equilibrium pattern would fail to be reestablished? By hypothesis, my sole contribution is to have it established one second sooner than otherwise. Now even a second counts: and after crops fail, society should even in the first second begin to reduce its consumption of grain. The worth of this one-second's lead time to society is perhaps $5, and if we for the sake of the argument accept a Clarkian naive-productivity theory of ethical deservingness, we might say I truly deserve $5. Actually, however, I get a fortune. My quickness enables me to reap the income effects of price changes without regard to whether I alone am uniquely capable of producing or signalling the desired substitution effects.[15]

Free competition among speculators would not necessarily lead to an optimal social result. Nor would making the market still more competitive—that is, increasing the number of speculators—lead to the disappearance of speculative profits in a world where speculators do not enter the industry with "omniscient" crystal balls. Indeed, what the theoretical analysis led to was a desire for a larger supply of "omniscience" —or knowledge, even from a central or governmental source —and a reduction in uncertainty. The pragmatic policy conclusion to which this reasoning led was to try to improve crop forecasting, speed up the communication of such information to farmers, and reduce the disturbances resulting from risk and uncertainty. This conclusion is opposite to the one reached by laissez-faire reasoning á la Hayek, and it could yield an improvement in social welfare. Though the crop forecasting and other actions of the Department of Agriculture are sure to be imperfect, the question is whether society would be better off with *no* central source of forecasting and a market left entirely to speculators.

What at first seems like a rather narrow technical argument thus widens out into a broader philosophical justification of the "mixed economy," in which both individual economic activities and government information gathering, communication, and planning have their role in enhancing the social welfare. In launching his critique of traditional defenses of speculation, Samuelson did not mean to deny any role to speculative markets. Having investigated the contribution to economic efficiency of commodity markets, he now concludes that "they aren't just casinos where portly gentlemen with small advantages in arithmetic take money away from everybody else, though they do have that aspect." They serve society by facilitating the production and distribution process and by relieving the burden of risk from those who cannot or do not choose to bear it.

For other purposes—such as protecting the nation from blackmail by a cartel of oil-exporting countries—he would have the government intervene in the market as buyer or seller, to bring down a monopoly price and safeguard American consumers' interests, and possibly national security. But he is under no illusion that the government will play its role just right. "I am anything but sanguine," he says, "that governments are very good at storing at the right time and releasing stocks at the right time. The political pressures are fantastic not to release." However, the real issue is not whether the government can be expected to act optimally but whether, given the real conditions in the marketplace (or the actions of other governments), a better outcome will result if the U.S. government does nothing at all.

He simply does not believe that private markets can resolve certain types of important problems, especially those loaded with "externalities"—that is, the activities of private individuals or corporations that have wider consequences for society, including benefits (such as improved health, education, or community development) or costs (such as pollution and environmental degradation). Samuelson believes that *knowledge*, though a scarce resource, is heavily freighted with externalities of great potential benefit to society. He sees no reason why one should presume that the functioning of private markets should lead to an optimal distribution of knowledge. Knowledge is a peculiar good, different from virtually every other. In the case of ordinary economic goods, such as bread, when one person gives up the good, he has less of it. With knowledge, when its possessor gives some of it to another, he still has as much knowledge as he had in the first place. Though it may be to the economic advantage of the original possessor to hoard his knowledge, society might be worse off.

Yet society wants knowledge to be increased, and that ("as every patent lawyer knows," says Samuelson) costs money. Creating a loaf of bread also costs money. But the loaf of bread belongs to the baker, and he will go on producing bread up to the point where its benefit to society (as measured by the price it brings in the market, where people buy voluntarily) is equal to its pecuniary and social cost, especially if there is competition among many bakers. Again, knowledge is different: the producer of knowledge knows in advance that others will imitate his discovery. He also knows that society will benefit by the full amount of the uses of his discovery everywhere. But he knows that if he is not careful, he himself may get little out of it, since the pecuniary benefits will revert fully to the creator of knowledge only under certain special (usually noncompetitive) conditions. As the individual creator pursues his self-interest, no invisible hand will lead him to the social optimum.[16]

It is difficult to say what *will* yield the social optimum because both the *production* and *distribution* of knowledge are involved. Joseph Schumpeter concluded that perfect competition would stifle dynamic progress; hence he justified private monopolies in knowledge, via patents and copyrights or simply by secrecy and exclusive possession. But Wassily Leontief goes the other way, arguing that, whatever the method of financing the discovery of new knowledge might be,

> the economic benefits of scientific and industrial research can be exploited fully only if no one, no one at all, is prevented from using its results by the price which he has to pay to do so. . . . In an era in which economic progress depends so much on scientific research, such chronic underemployment of technical knowledge might have, in the long run, an even more deleterious effect on the rate of economic growth than idle capital or unemployed labor.[17]

The implication of Leontief's argument is that the government itself should pay for the creation of new knowledge, as necessary to ensure its full use, rather than permit its private monopolistic restriction. On the other side, the believer in laissez-faire argues against both public subsidy and private patents and monopolies.

For Samuelson, the issue cannot be resolved theoretically or by abstract logic. The policy problem becomes completely pragmatic: the solution requires the careful weighing of the harm that would be done by monopoly restrictionism after the new knowledge was created, against the inducement to creative change that would result from providing strong pecuniary incentives to investment in new knowledge.

In any case, private enterprise, even with monopoly protection, will not invest in all the types of new knowledge that it might be desirable for society to have—because there is no existing market for certain types of knowledge, because it might take a great many years before there was a pecuniary payoff from such knowledge, because the risks of finding new knowledge or usable technology are too great, or because the costs to be incurred are beyond the resources of any firm or interested group of firms. Thus society, through government, must invest collectively in the quest for certain kinds of new knowledge that might serve the social welfare, if the effort is to be made at all. This could apply to fields ranging from the study of simple microorganisms to experiments with nuclear fusion and solar energy.

VII

SAMUELSON HAS, from the beginning, pursued that will-o'-the-wisp (or Holy Grail) of the *social welfare*, which is indeed the ancient and most fundamental goal of economists. The grand, underlying theme of Adam Smith's *Wealth of Nations* was his belief that the pursuit of individual self-interest and the attainment of the social welfare are harmonious: ". . . [no] individual . . . intends to promote the public interest . . . he intends only his own gain, and he is in this, as in many other cases, led by an invisible hand to promote an end which was no part of his intention."

Drawing on the work of his Harvard colleague Abram Bergson, Samuelson showed that Smith's basic tenet—the core of the classical and contemporary case for free enterprise and against state interference—rests on the assumption of *perfect competition*. The libertarian case, therefore, stands or falls on the degree to which economic reality corresponds to perfect competition. But one of the basic intellectual revolutions of twentieth-century economics—just before the Keynesian revolution—was the demonstration in 1933 by Edwin H. Chamberlin of Harvard University and Joan Robinson of Cambridge University that in modern industrial economies, the normal case is that of *monopolistic competition* (Chamberlin's term) or *imperfect competition* (Robinson's term), both of which mean the domination of major markets by a small number of firms that take account of each other's actions. The entire libertarian case for laissez-faire therefore seemed (and still seems) to Samuelson to rest on feeble assumptions. "What libertarians have in common," he says, "is the hope that departures from perfect competition are not too extreme in our society." He notes that Milton Friedman has apparently come full circle to the view that public regulation of monopoly is a greater evil than letting well enough alone. However, some other champions of the classical liberal view still put considerable emphasis on anti-trust action or other measures to preserve competitive markets.

As a description of the actual state of the American economy, the theory of monopolistic competition was extremely inconvenient to the University of Chicago's libertarian economists, who sought to dismiss it as sloppy or dangerous in its policy implications. Frank Knight, the dean of the Chicago School, announced at a Princeton meeting in 1950 that "if there is anything I can't stand, it's a Keynesian and a believer in monopolistic competition." Samuelson, who was there, needled Knight by asking, "What about believers in the use of mathematics in economic analysis, Frank?" Knight replied that he couldn't stand them either, and Samuelson knew that the negative Chicago verdict fit him to a "T."[18]

Milton Friedman was unwilling to let the threat which the reality of monopolistic competition posed to laissez-faire doctrine be neglected, lest it grow like a tumor. Friedman sought to excise it by his methodological excursion into "positivistic economics." He maintained that the assumptions of a theory need not be an exact description of reality in order for the theory to be valid and useful. Economics should proceed "positively," reasoning from general assumptions to specific conclusions; if these conclusions proved to be correct, one could presume that the assumptions were satisfactory. Samuelson charged that Friedman had given this seemingly reasonable proposition an "F-twist"—the contention that "a theory is vindicable if [some of] its consequences are empirically valid to a useful degree of approximation; the [empirical] unrealism of the theory 'itself,' or of its 'assumptions,' is quite irrelevant to its validity and worth."[19]

Samuelson branded as nonsense the argument that factual inaccuracy could be anything but a demerit for a theory or hypothesis. Some inaccuracies, admittedly, are worse than others, but that only means that some sins are worse than others, "not that a sin is a merit or that a small sin is equivalent to a zero sin."[20]

While regarding Friedman's "crypto-positivism" as of no moment to a philosopher or scientist, Samuelson found the "F-twist" of interest to economists because of its motivation, which he said was to help the case for the perfectly competitive laissez-faire model in economic theory. This has been under attack from outside the economics profession by humanists, nationalists, Romantics, and others for over a century, and from inside the profession since the monopolistic-competition revolution of the early 1930s. The "F-twist" also helped safeguard the traditionalists' hypothesis that the aim of business firms was simply to maximize profits, a hypothesis which Samuelson called a mixture of "truism, truth, and untruth." Friedman's crypto-positivism was designed to serve Friedman's and the Chicago School's own "nonpositivistic" —that is, normative—purpose, which was the defense of laissez-faire and private enterprise, as well as the effort to bail out a failing economic model in which some economists had invested all their intellectual capital.

The debate over positivism highlighted an important issue that is frustrating both to economists and to the public at large: the fact that economics tends to rely on very general models, theories, or paradigms in an effort to give coherent explanations to a wide variety of imperfectly observed or measured phenomena. These economic paradigms involve a measure of hunch or faith, a kind of artistic simplification of reality. While insisting that the realism of a theory's assumptions is a relevant question, Samuelson agrees that "of course, your assumptions can be idealized—if you tried to explain everything down to the last decimal point, you would have a theory that would totter of its own weight." It is not just the economist whose theories describe reality only up to a certain degree of approximation. "Every theory," says Samuelson, "and this was true long before quantum-theory days, has a problem of errors in measurement, the scatter around laws. In higher approximations, you learn something about that scatter." But that is very different from saying that a theory is all the better for its imperfections.

New perceptions of reality—or of new and evolving realities—require new paradigms, as Thomas Kuhn suggested in *The Structure of Scientific Revolutions.*[21] The paradigm of monopolistic or imperfect competition has relevance to the marketing behavior of large corporations, to output trends in particular industries, to problems of unemployment and industrial mobility, and to many other processes in an advanced industrial economy. Samuelson concedes that perhaps the paradigm of perfect competition, like the paradigm of a frictionless model, still provides some hypotheses that are "robust" on particular topics or for particular markets. But if you are trying to explain why a dying industry keeps the same work force and operates only a minor fraction of the year, the deviations from the perfect-competition model are the essence of the problem, and one should turn to a new model rather than cling to an old one that only tells you what is not happening but "ought to."

Unlike pure mathematics, economics does not seek to build a beautiful, coherent, logical structure but to depict reality faithfully, to ferret out and capture the regularities of economic phenomena and of people's economic behavior. "Economists are often envied," says Samuelson, "because a subset of what they should be dealing with, and for many of them the whole set of what they do deal with, often does have a good deal of regularity. People in the other social sciences envy us for having the easy questions to answer." But he doesn't think a good economist can be only an economist. He finds it sad that when working on problems which overlap with other disciplines, the economist has to become an amateur anthropologist or sociologist or psychologist. "How nice it would be," he says, "if in fact you could have a division of labor, and one man would hold the tumbler and the other man would do the somersaults. We are all looking for that other fellow, I think, as we look for that good broker—the man in the other discipline who meshes with our interests."

He is skeptical about the ability of economics as a discipline to invade and make useful contributions to unfamiliar social fields. Economists have recently been writing on social discrimination, crime, alienation, education, and other sociological and psychological topics. "I have only a modest enthusiasm for that literature," says Samuelson, "because it seems to me that a good deal of it is simply stating things in economic terms, restating ideas so that they seem shocking to the noneconomist and flattering to the economist."

Samuelson thinks the economist often deals in rather far-fetched analogies when he wanders far afield. "To understand the demographic revolution which took place after 1939, and why the wheel again turned back after 1957," he argues, "you don't gain very much by an elaborate analogy between children and Cadillacs and inferior goods. Families have fewer children, but children are not inferior goods, which that literature works hard to show."

The particular formal structures offered by economists to explain population trends seem to him arid and empty. "If I wanted to understand demographic change," he says, "the test for me would not be allegedly to explain what has happened but to be prepared for what will happen in the years to come." He suggests that instead of analyzing the demand for children by analogy with the demand for inferior goods, economists "could learn as much by applying Seventh Avenue analogies of skirt lengths and fashions, because the contamination of tastes is more important than the structure of tastes at any given time."

He illustrates from his own family's experience; he and Marion Samuelson are celebrated among economists for having had six children—first one girl, then another; then a boy; and then boy triplets. Treating his own family as social data, Samuelson says, "When my wife and I were married, we didn't have children for eight years. And six or more of those years were by choice. It was not fashionable to have children then. But the wheel turned. It got to the point where if you weren't married, people thought you were a homosexual, and that was bad. Now the wheel is turning again."

But he insists he has nothing useful to say about how procreational phenomena may change again; he has no expert knowledge on how the social curves will shift. "Only people who are close to fashion are worth listening to on fashion." Similarly, if you want to know where constitutional law is heading, people close to the Supreme Court are worth listening to. "The principle," he concludes, "is the Frank Ramsey test of bet-ability: knowledgeable people go wrong, but you can bet they will go wrong less often than others."

This way of thinking underlies Samuelson's cautious approach to economic forecasting. He tells how Professor Tjalling Koopmans of Yale, a meticulous mathematician and econometrician and later a winner of the Nobel Prize in Economics, came to Samuelson in 1955 for his views on the progress being made in economic forecasting and the potentials of the field. Samuelson told him that economists were forecasting better, and more people were forecasting well than ever before, because of more and better information and better forecasting models. "But," Samuelson added, "I don't think we're converging on more and more accuracy. I think we're converging toward a kind of irreducible Brownian motion, or you could give it a highfalutin name like the Heisenberg indeterminacy principle, because God Almighty hasn't made up His mind as to what business investment is going to be next year, so how can we read tea leaves and find out what it is?"

Samuelson thinks this observation has proved to be only too true, and he still sees "no sign of anything, under present methodology, converging toward great accuracy." This conclusion, he says, "may sound humble, but it is really arrogant of me—since I contend that it is my knowledge of the ignorance that is knowledge." Samuelson forecasts only in the most gingerly fashion himself; what he really does in his outlook articles for the *Financial Times* of London or *Nihon Keizai Shimbun* of Tokyo is to comment on the forecasts of others.

He believes in guessing. But the guess should have a conscious or unconscious empirical basis. "Arthur Burns' finest hour," he says, "was the 1955 recovery. He forecast higher than all his staff. I asked one of his staff 'How come?' and he said, 'Well, Burns has reasons we will never know.'" But Samuelson thinks Burns's real reason was that the General Motors cars of that period had met a resonant response: "You could tell it from the cigaret smoke in the salesrooms early in the autumn of '54." He thinks of science as uncommonly common sense, including a willingness to recognize what you can't know.

When he is called for advice by the U.S. Treasury or the Federal Reserve Bank of Boston, as he sometimes is, Samuelson tells his clients that they ought to go beyond taxonomy to what's the best present guess or set of guesses. He forecasts within wide ranges, in which he expects two-thirds or four-fifths of the observations of others to fall, and tells policymakers that they should base their policies on such wide bandwidth assumptions. In his own stock market investing, he always widens the odds well beyond what any broker or adviser tells him.

VIII

THUS, both a stock market investor and a social reformer, Samuelson tries to keep positive economics separate from normative economics. Following the line of Lionel Robbins's *The Nature and Significance of Economics* (1930), Samuelson says that it is essential to realize that all normative (prescriptive) statements are in a different grammatical mood from positivistic statements: they are *ought* statements, not *is* or *are* statements. "It may be a positivistic fact," he says, "that in a society with great evenness in the distribution of income, there is less rickets and the life-expectancy dispersion among different categories of the population is less. But to say that this income distribution *ought* to be achieved is a statement in a wholly different mood. That seems to me to be quite important. You can't have normative economics without norms, without ethics, without prejudices."

But are ethical norms purely subjective, or can they be firmly established? Samuelson doesn't resolve that question in an ultimate sense. Instead, he tries to clarify the way people can increase their society's subjective welfare—that is, improve the welfare of the whole community according to the subjective standards of the individuals who make up the community. He contends that people can, acting individually and together, establish what are to be their norms. "I, for example, all my adult life, including the present time," he says, "have had a strong egalitarian prejudice." Yet he recognizes that equality is not the only norm; real output and income count as well. And egalitarianism may come at a price, even to the poor. If it leads to a loss of total effort and production in the society, there will be smaller total output and hence smaller absolute shares for all. The social welfare problem, he emphasizes, needs to be conceived and solved dynamically, not on the assumption of a fixed set of resources and desires.

He finds it "naïve" for Professor James Coleman, the sociologist whose research on racial integration in education provided much of the impetus behind busing, now to discover that when you try to integrate the schools there are disturbing secondary consequences. "That," says Samuelson, "is another move on the chessboard. It may be a move which we do not like, such as people leaving the inner city so that you may end up with more segregation. That's precisely what a good chess player has to take into account."

Samuelson believes that if there is to be long-term improvement in social welfare, ethical norms must be established. However, these norms have to come "from outside of anything established by possession of a Ph. D. in economics, or by virtue of being a Ph. D. in any field, or by anything you can measure, such as the speed of light or aggregate demand."

How are ethical norms established?

The answer to that question, when it is interpreted positivistically, is relatively simple: ethical norms are established by sacred texts and a priesthood; or by a dictator or the dominant bureaucracy; or by the principle of every man or every clan for himself or itself; or by an evolutionary process of bargaining among individuals and groups—for instance, as religious toleration developed in the American colonies as a consequence of "the common interest of the numerous sects in preventing domination by any of the others."[22]

But the questions of how ethical norms *ought* to be established and what those norms *should* be are quite different, and far more difficult to answer. Adam Smith, as we have seen, offered his intuitive thesis that the good society would emerge from the pursuit of self-interest, under conditions of free competition, with no powerful political authority to determine the outcome. Several nineteenth-century economists, including Thomas Malthus, John Stuart Mill, Léon Walras, and Alfred Marshall, recognized that if free competition were to eventuate in a "just" state, the initital distribution of wealth had to be just; if the initial conditions were unjust, so would be the social outcome, for the marketplace can be as much a source of coercion as the state. Malthus understood this coercion of the market very well, noting that Fate dealt a cruel hand of cards to the worker's child and a favorable hand to the well-born. Mill said that a lot more than Fate was involved; the rights of private property, enforced by the law, counted heavily against the poor; the injustice of property distribution today could condemn future generations to misery and starvation, which the market would not correct.

Samuelson believes that there can be injustice in both market and nonmarket forms of coercion, and indeed the effects may be the same. He offers, as a test, this simple statement: "I am kept from attending college. My family is ———." In different societies, with different ethical norms, the blank might be filled in with *black, bourgeois, Jewish,* or *poor*. Samuelson agrees with Anatole France's ironic tribute to equal justice under law: "How majestic is the equality of the Law, which permits both rich and poor alike to sleep under the bridges at night."[23]

Yet Samuelson is not necessarily against all forms of coercion, even state coercion, if a worthy ethical norm requires it. In this, he differs once more with Milton Friedman, who argues that it is better for one who deplores racial discrimination to try to persuade others against it than to do nothing at all; but, failing to persuade, there should be no coercion, even democratically arrived at, because a general precept against discrimination would be arbitrary and gratuitous.[24]

Samuelson says he can prove the "absurdity" of this contention, concealed by its abstract form, with a single counter-example. Samuelson rephrases Friedman's proposition in this way:

> If free men follow Practice X that you and some others regard as bad, it is wrong in principle to coerce them out of that Practice X; in principle, all you can do is try to persuade them out of their ways by "free discussion."[25]

Samuelson's refutation is as follows: Substitute for the term "Practice X" the phrase "killing by gas of 5 million suitably specified humans." "Who will agree with the precept now?" he asks. Samuelson answers his own question by saying that only two types will possibly agree: (1) those so naïve as to think that persuasion will possibly keep Hitlers from cremating millions, and (2) those who think the status quo achievable by what can be persuaded is "pretty comfortable, if not perfect." He excludes a third type: those who simply accept an axiom without regard to its consequences or who don't understand what its consequences are.

Samuelson's dramatic "exception" to Friedman's general argument against political coercion (even in a worthy cause) signifies his own conviction that moral issues commonly involve a reconciliation among "goods," such as freedom and justice. Samuelson believes that life consists of minimizing multiple evils and maximizing multiple goals, by compromise when possible, while preserving as wide a role as possible for individual freedom—but not disdaining to use political means if they are necessary to achieve some social good.

Samuelson's deep concern for fundamental principles of human justice coexists with an equally profound desire for every individual to choose his or her own values. He dislikes trying to impose his own ideas on others. "On the stump," he says, "I am always being asked this kind of question: 'Suppose you had unlimited power, what would you do?' That's the kind of question I try to duck. But when I feel the responsibility to testify, as before a congressional committee, I always try to say, 'Well, the answer I give you is based in part on my scientific knowledge (I don't put it in such a stuffy way as that) and in part is a value judgment.' "

He doesn't like to lay down the law even to his own children. "You could guess I am a fairly permissive father; I don't have a strong opinion on exactly how a youngster should be brought up or what he or she should choose." He finds advice giving all too cheap and easy, on personal matters as on matters of state.

He sees social, political and economic life as a continuous process of conflict resolution. But not every conflict can be resolved freely, peacefully, or in time. He acknowledges that he "could not expect Congresswoman Shirley Chisholm, representing the Bedford-Stuyvesant constituency in Brooklyn, and a congressman from St. Petersburg, Florida, representing a group of old, retired, middle-class people, to reach a compromise on the rate of inflation versus the rate of unemployment, assuming that actions to curb one will aggravate the other." Should the burden of unemployment be put upon those whose employability is most volatile in response to changes in effective demand, such as the blacks of Bedford-Stuyvesant? Or should the burden of inflation be put on the aged of St. Petersburg? "There is no way," says Samuelson, "that those people can agree on the same thing." Besides the immediate clash of values and interests, there are intertemporal differences: "Some of those people are going to be dead soon. Some will have concern for posterity, and others will say, like Louis XV—Louis XVI wasn't smart enough— '*après moi, le déluge.*' "

Such conflicts cannot be resolved, he claims, by facts or logic; there is no way to settle the conflict without making value judgments. In that respect, Samuelson thinks he has walked some distance along the road with Gunnar Myrdal and has discovered that it is a lot harder to separate value judgment from positivistic, observable fact than he once thought. "If you were to do a variorum study of my textbook in all ten editions," he says, "you would see changes in the wording on these matters."

To be healthy and stable, he suggests, a free society—one in which individuals are free to hold conflicting ideas and pursue ends of their own choosing—must somehow arrive at shared ethical values. Shared values are those that most people believe to be in both the society's and their own interests. Those who violate such ethical principles usually recognize their deviance and experience guilt.

Sometimes the entire society benefits from a shared value and a government policy that brings about change in the name of that value. "There are some things that are done so stupidly in a society," says Samuelson, "that everyone can gain. Once the overall situation is improved, there will be bribes for everybody."

But much of the time, he fears, it is "brother against brother—and then you simply have to appeal to shared ethics, since some will gain and others will lose." (In other words, if benefiting the poor actually costs the rich something, one must appeal to the rich and the middle class in terms of some shared ethical principle, such as love, charity, equality, or need.) Samuelson thinks it best if the members of a society reach shared ethical principles voluntarily through their own reasoning and convictions. He finds it "really quite wrong—this is my personal-guilt philosophy talking—to force any of my idiosyncratic ethical views on an undeceiving larger public, just because I am a pretty good debater and a pretty resourceful knower of facts about economics."

He declines the role of rabbi or prophet. "Kenneth Boulding," he says, "is a mystic. But I have no real mother's milk religious background." His candor on his freedom from religious influence is refreshing, if a bit startling: "I was at Yale as a Whyte Fellow at Calhoun College, and the master of Calhoun that year, a professor of Old Testament, said to me, 'Oh, you must have been steeped in the Bible.' I said, 'Why would you say that?' And he said, 'Well, your every quotation is from the Bible.' I said, 'Well, in that case, I am steeped in *Poor Richard's Almanack* and Shakespeare—never read the Bible through, just a literary acquaintance with it."

I X

SAMUELSON'S this-worldly and democratic social philosophy forms an organic whole with his principles of economic policy. He is a philosopher of the mixed economy, blending private and public sectors, market and political decisions, individual liberties and social responsibilities, personal and communal welfare. For him, as for the biologist Lewis Thomas,[26] man is a social creature. "Not even an individual's perfections," says Samuelson, "are his own; like his imperfections, they are group made. We entered a world we never made, and leave one we did not unmake. Carry the notion of the individual to its limit and you get a monstrosity, just as you do if you carry the notion of a group to its limit."

His social and political ideas have been sharpened by his work in economic theory. His constant balancing of alternatives, his calculation of trade-offs, his search for "catenary saddle-points" and equilibria and maxima and minima and optima—all these have shaped his political and social thinking. His mental models are those of economics.

For him, mathematics is literally *language* (not merely *a* language, as Willard Gibbs said) and great fun, to boot. He says: "I think having children is the biggest kick in life, and the modern generation doesn't see it that way at all, but aside from that, the greatest pleasure I have—the greatest personal pleasure—is in the puzzle-solving aspect of economics, the mathematical work. But in the end, the puzzle is a much better puzzle if it isn't just a puzzle, if it has relevance to real-world problems."

Samuelson's lifelong aim has been to put his mathematical gifts to the service of humanity. He does not think he has been practicing a bloodless art, or preaching an apologia for the status quo. He wound up his Nobel address in Stockholm by saying, "An American economist of two generations ago, H. J. Davenport, who was the best friend Thorstein Veblen ever had—Veblen actually lived for a time in Davenport's coal cellar—once said: 'There is no reason why theoretical economics should be a monopoly of reactionaries.' All my life, I have tried to take this warning to heart, and I dare call it to your favorable attention."[27]

The miracle is that the abstractions of economics sometimes do have a powerful impact on society and help to solve excruciating problems that afflict many millions of people all over the world. Samuelson has been a pioneer in the reconstruction of economics into a coherent and orderly discipline. He has done more than anyone of his time to disseminate its insights and to influence government policy at the highest level. With Keynes and the other outstanding economists of his time, Samuelson has helped to lift the scourge of mass unemployment from the capitalist nations and to prevent the danger of systemic breakdown. With all the unsolved problems that remain—ranging from inflation to environmental degradation—the enormous contribution of economic scholars to economic stability should not be underrated or taken for granted. It provides hope that free societies can survive and ultimately solve the other problems of a new era.

But Samuelson, dedicated economist that he is, seems to sense that economic science is inadequate to the tasks which lie ahead. For many of society's problems, he admits that "all there is usually time for is a quick and dirty solution." The

Paul Anthony Samuelson

best advisers, he says, have simple models—and courage. Genius consists in finding the models, such as "the market" or "the macroeconomy," that correspond closely to aspects of reality. Is society in need of new models? Must they come from outside economics, a science rigidified by its past triumphs?

Milton Friedman

Prophet of the
Old-Time Religion

"Capitalism and altruism are incompatible; they are philosophical opposites; they cannot co-exist in the same man or in the same society."

—*Ayn Rand*

"There is no such thing as a free lunch."

—Anon.; also book title,
Milton Friedman

I

ADAM SMITH is generally hailed as the father of modern economics, and Milton Friedman as his most distinguished spiritual son.

Building on deist theology, according to which God, after setting the universe in motion, abandoned it and assumed no control over natural forces, Smith set out to demonstrate that an economy functions best when it operates on the basis of the forces of self-interest and competition. Such an economic system, he contended, has no need of altruism; on the contrary, altruism—a rare human trait—is less reliable and powerful a force than self-interest in creating social wealth and serving the needs of others. However, the natural force of self-interest, which, said Smith, "comes with us from the womb and never leaves us till we go into the grave," can serve the common good only if government or some other powerful monopoly does not interfere with the voluntary actions of individuals and businesses. The injunction of Smith to the state was let it happen—*laissez faire*—let self-seeking businessmen and individuals pursue their own interests without interference from government.

The remarkable contribution of Milton Friedman, in the wake of the Great Depression of the 1930s that had threatened the very existence of the free-enterprise system, was to show how much life was left in the old-time Smithian creed—and how many ingenious applications could be devised to harness self-interest and the forces of the marketplace to the solution of social problems.

Friedman became an ardent crusader for capitalism and

economic freedom at a time when most of his leading con-
temporaries in economics were seekng ways to use government to improve the performance of the economy and to
advance social welfare. Friedman's devotion to the laissez-
faire philosophy, and his skill at arguing its continued relevance to a highly industrialized and organized society, made
him the most celebrated economic conservative in America, a
hero to businessmen in the United States and many other
countries. But he was not simply Adam Smith redivivus.
Smith, wary of business interests and their tendency to form
monopolies against the consuming public, wrote at a time
when the industrial class was seeking to rise against the resistance of the dominant feudal authority; Friedman came as
the defender of a business class already in the saddle.

Friedman's own origins were lowly. He was born in Brooklyn, New York, in 1912, the son of poor Jewish immigrants.
Before he was born, his mother had worked as a seamstress in
a New York sweatshop, and his father dealt in wholesale dry
goods. A year after Milton's birth, the Friedmans moved
across the Hudson River to Rahway, New Jersey, where they
set up a retail dry-goods store. His mother ran the store, while
his father continued to go into New York to run his wholesale
business. His father died when Milton was fifteen years old,
leaving his mother with very little money to pay for their
son's education. Milton had received a religious upbringing,
but by the time he was thirteen, he had decided that religion
was nonsense, and never had anything further to do with
it.

His greatest aptitude was for mathematics. He was a hard-
working student who was not only brighter but younger than
most of his classmates. He graduated from Rahway High
School before his sixteenth birthday and won a New Jersey
state scholarship to attend nearby Rutgers University in New
Brunswick. He supported himself at Rutgers by waiting on
tables and clerking in a department store. For fun he worked

as a copy editor on the student newspaper. He was required to take two years in the Reserve Officers Training Corps, which, he says, gave him a lifelong dislike of compulsory military training.

As an undergraduate at Rutgers, he cast about for some practical way of capitalizing on his aptitude for mathematics. He decided to become an actuary, and passed the actuarial examinations while he was still in college. But by the time he left Rutgers, he had discovered—and fallen in love with—economics.

Two young men, Arthur F. Burns and Homer Jones, were responsible for introducing Friedman to economics. Both were teaching at Rutgers while completing their doctoral work at other institutions—Burns at Columbia University and the National Bureau of Economic Research, and Jones at the University of Chicago. The influences of his instructors and their institutions were seemingly disparate. Chicago was a center of work in economic theory and the home of a very conservative economic ideology. But Columbia—or, more particularly, the National Bureau of Economic Research, under the dominance of Wesley C. Mitchell, its principal founder, president, and research director—stressed a would-be scientific empiricism, a search for "the facts" rather than abstract theories of economic behavior.

Friedman later sought to show that the difference between a search for facts and a search for theories was more apparent than real. In a commemorative volume about Mitchell, Friedman wrote:

> Wesley C. Mitchell is generally considered primarily an empirical scientist rather than a theorist. In my opinion, this judgment is valid; yet it can easily be misunderstood. The ultimate goal of science in any field is a theory—an integrated "explanation" of observed phenomena that can be used to make valid predictions about phenomena not yet observed. Many kinds of work can contribute to this ultimate goal and

are essential for its attainment: the collection of observations about the phenomena in question; the organization and arrangement of observations and the extraction of empirical generalizations from them; the development of improved methods of measuring or analyzing observations; the formulation of partial or complete theories to integrate existing evidence.

In this sense, Wesley Mitchell's empirical work is itself a contribution to economic theory—and a contribution of the first magnitude. . . . There is of course no sharp line between the empirical scientist and the theorist—we are dealing with a continuum, with mixtures in all proportions, not with a dichotomy. "The most reckless and treacherous of all theorists is he who professes to let facts and figures speak for themselves." And, one might add, the most reckless and treacherous of all empirical workers is he who formulates theories to explain observations that are the product of careless and inaccurate empirical work.[1]

Mitchell's protégé Arthur Burns was essentially an empiricist. Burns's nearly completed doctoral dissertation on production trends in the United States, which he was writing under Mitchell, served as the basis of a seminar he gave at Rutgers in Friedman's senior year. From that seminar Friedman gained an appreciation of the role of "value-free" statistical research.

Meanwhile, Homer Jones was filling his star pupil's head with dreams of Chicago, and encouraged and supported Friedman's application for a graduate scholarship there. When Friedman graduated from Rutgers in 1932, at the very bottom of the Great Depression, with a joint major in economics and mathematics, he received offers of graduate scholarships from Brown in mathematics and Chicago in economics. He chose Chicago.

Chicago was an exciting place for a budding economist in the 1930s. Frank Knight was the philosophical heavyweight among the conservatives in its economics department. He held that the state, even when it sought to do good, was far more likely to do evil. Knight was deeply pessimistic about the human race and its pretensions of altruism; he thought the dominant trait of mankind was greed, and he thought it essential to build an economic system which recognized that fact, harnessed greed for useful purposes, and thereby dispersed it among many competing centers of power. This was the pervasive Chicago philosophy.

But Chicago was much more than the home or museum of a Smithian ideology. Knight himself was a gifted theorist; his book, *Risk, Uncertainty and Profit* (1921) had laid the foundations of the modern theory of the firm. Aaron Director (Paul Samuelson's first economics teacher) was a mordant critic of governmental and social institutions. Jacob Viner was one of the greatest theorists and historians of economic thought who ever lived; Friedman calls Viner's theory course "an absolute eye-opener—one of the two or three greatest intellectual experiences of my life."

Ideologically, Chicago's economics department was by no means all of one piece. Henry Simons, a brilliant and original monetary theorist, was a determined foe of big business and monopoly; and where the market could not work, he was willing to see government socialize. One of the great stars of the Chicago department was Paul Douglas, who had grown up on a Maine farm and who joined his twentieth-century liberalism and political reformism to a remarkable gift for economic theory. Douglas, later to become an influential Democratic senator from Illinois, made outstanding contributions to the theory of marginal productivity, the theory of wages, and the analysis of factors underlying production and economic growth.

Chicago respected empirical work as well as theorizing. When Friedman arrived there, he was made an assistant to Henry Schultz, who was doing pioneering work in the statistical derivation of demand curves.

For Friedman, what Chicago stood for was "theory with empirical content." The quest for valid theory was what economics was all about. Although Knight was the major philosophical influence on Friedman, the young scholar did not immediately become a "true believer." He does not remember Knight as having been a good teacher. "Two-thirds of Knight's students," says Friedman, "got nothing out of his courses, and the remaining third got nothing two-thirds of the time—but what was left was magnificent."

Knight attracted many disciples, but Friedman now says that his influence on them was usually not healthy. Knight's acute critical faculties, combined with his growing sense of the limitations of economic analysis, tended to stifle those who looked to him for guidance and support. According to Friedman, Knight's natural "contrariness" often stifled scholarly productivity. His favorite sayings were "All simple statements are wrong" and "No statement is so absurd that it doesn't contain some truth."

Friedman appears to have absorbed both these Knightian dicta into his own being. He delights in making short, dramatic pronouncements which he quickly acknowledges as partially wrong while insisting on their basic truth. This style and strategy make him a formidable and ingenious debater, and a tough opponent to catch. This sometimes infuriates his opponents. Joan Robinson, after a frustrating encounter at Chicago, called Friedman a "paper tiger." Paul Samuelson says, "Now I don't think Milton is a charlatan. . . . He believes what he says at any time he says it. But he also has a very healthy respect for his audience. If you are a yokel, he gives you a hokum answer. If he is giving his presidential address, he states it more guardedly and more carefully. . . . It is simply a matter of the style of the person."

I I

FOR FINANCIAL REASONS, Friedman had to leave Chicago after the 1932–1933 academic year. A job waiting on tables had insufficiently supplemented his $300 tuition scholarship. He had been forced—for the only time in his life—to go into debt.

When Columbia University promised him a $1,500 fellowship, he jumped at the lavish offer. After paying off his debts in Chicago, he moved to New York, where he spent the next year completing preliminary work for his doctorate, frequenting the Broadway theater, and living better than he ever had before. At Columbia he was most influenced by Mitchell and Harold Hotelling. Hotelling was an abstract mathematician with an uncanny instinct for sniffing out problems in statistics and economics which were only later to be recognized as important. One of his articles in the early 1930s, for example, dealt with the economics of exhaustible resources. Another was a pioneering exploration of the effects of taxation and railroad and public utility rates on the general welfare; it opened the way to a more scientific evaluation of social costs and benefits. As Paul Samuelson has noted, it was at Columbia, before World War II, that "Hotelling became the Mecca toward whom the best young students of economics and mathematical statistics turned."[2] Friedman was one of these.

Wesley Mitchell tempted Friedman with institutional economics, that peculiarly American counterrevolution against economics as an abstract, inferential science, but did not capture him. A student of Thorstein Veblen and John Dewey, Mitchell was highly skeptical of the effectiveness of deductive reasoning (proceeding from the general to the specific) in the social sciences. Mitchell did not believe that rational calculation of interest underlay all economic decisions. For instance, he felt that it was a mistake to suppose that consumers guided their course by ratiocination. Because their behavior was not necessarily rational, it had to be observed rather than deduced from abstract principles. Mitchell took economic theorists to task for ignoring this. His own work was largely devoted to the detailed study of business cycles. He accumulated vast quantities of data, often in the form of statistical time series; despite Friedman's later effort to show that Mitchell's work was implicitly theoretical, it actually devoted little attention to theoretical analysis.

Friedman took two courses from Mitchell at Columbia, one in the history of economic thought and the other in business cycles. The first he says he found uninspiring. But the second, Mitchell's true passion, was more illuminating and exciting. Friedman says that no one exemplified the "scientific spirit" more for him than Mitchell.

In 1935, Friedman took a brief leave from academia to join the New Deal—and to work, of all things, in an agency concerned with long-range planning! The Industrial Section of the National Resources Committee was gearing up for a study of family consumption patterns, and Friedman took a job as a member of Hildegarde Kneeland's Consumption Research Staff. The study was designed to provide "factual information on the consumption aspects of the economy, against which to evaluate the many current proposals for economic recovery and expansion."[3] Friedman apparently had no compunction about participating in such a venture; he had the classical economist's respect for the free market, but not yet the libertarian's belief in it as the demiurge of all social and economic problems. Moreover, as far as the study went, policy was not his concern; his job was as a statistician, formulating methods and interpreting statistical results.

Although the study, *Consumer Expenditures in the United States*, was published in 1939, Friedman spent only two years in Washington. In 1937 he received an offer from Simon Kuznets—later to be awarded a Nobel Prize in Economic Science for his pioneering work on gross national product and economic growth—to come to work at the National Bureau of Economic Research in New York. When Friedman accepted and joined the Bureau, his immediate task was to take over from Kuznets a study of independent professional practice in the United States. The project, which was an examination of the incomes of physicians, dentists, lawyers, certified public accountants, and consulting engineers, became Friedman's doctoral dissertation.

Friedman prepared a preliminary study which the Bureau brought out in 1939. In the course of its preparation, Wesley Mitchell, as the Bureau's research director, took Friedman aside and criticized him severely for the poor quality of the writing. Kuznets—an immigrant to the United States in his teens—could be excused, said Mitchell, for writing badly, "but you don't have that excuse." From then on, Friedman always took pains with his prose; his simple and vigorous style later helped him to win converts to "Friedmanism."

Besides working on the study of professional incomes, Friedman also served as general secretary of the Bureau's Conference on Income and Wealth; he edited the second and third volumes of *Studies in Income and Wealth*. His own comments on one of these studies, "The Correction of Wealth and Income Estimates for Price Changes," by M. A. Copeland and E. M. Martin, are interesting for what they reveal about the young Friedman's intellectual proclivities. Copeland and Martin's paper was an attempt to assess real rather than purely monetary changes in income and wealth—"an attempt," as Friedman put it, "to get behind the monetary veil . . . and to measure changes in magnitudes that are considered in some sense more fundamental than value sums."[4] To Friedman this was a task of awesome complexity, a complexity of which he felt Copeland and Martin were largely unaware. As an example, he outlined the relevant considerations for discounting the effects of technological change:

> In order to obtain for any particular year the two figures needed—namely, actual "real output" and the "real output" that would have been produced had techniques remained unchanged—it is necessary to determine four things: first, the various combinations of output items that *could* have been produced with the resources available in the given year had techniques remained unchanged; second, the particular combination of output times what *would* have been produced; third, the "real output" that combination represents; fourth, the "real output" the combination actually produced represents. In order to arrive at the first it would be necessary to know the "production functions" corresponding to the techniques of the base year; to arrive at the second, it would theoretically be necessary to solve the equations of general equilibrium—not, note, the classical equations applicable under conditions of perfect competition, but those applicable to the real economy; to arrive at the third and fourth requires the solution of the problem of measuring "real output" not only for actually consumed baskets of goods but also for hypothetical ones.
>
> My purpose in stating the problem in this fashion is not, of course, to suggest any practical solution, but rather to indicate the complexity and difficulty of the problem, and the kind of knowledge required for an exact solution. The real problem, of course, is how, on the basis of observable data, to arrive at approximations to this exact solution that can be reasonably expected to be sufficiently close for the purposes for which they are intended.[5]

Friedman appeared to be suggesting either that such approximations were impossible to obtain or, at best, that the results would not be useful enough to justify the effort. He also took Copeland and Martin to task for what he considered their ideologically biased analysis. Implicit in their treatment was the doctrine that individuals ought to receive income proportionate to what they produce. Such ethical predilections, Friedman asserted, should be strictly divorced from scientific inquiry.

> Under a laissez-faire economy individuals may be able to obtain the value product attributable to their activities; but this is fundamentally different from saying that such a system of distribution is ethically desirable. "To each according to his abilities" may be the rule; "from each according to his abilities, to each according to his needs" may nevertheless be the ethical objective.[6]

Did this argument reflect a sympathy for Wesley Mitchell's social concerns—or merely Friedman's austere standards of scholarly objectivity? Whatever the answer, Copeland and Martin did not take kindly to Friedman's criticism; in their response they cited his use of "the somewhat high-sounding language of an absolutist philosophy."

I I I

IN 1938, Friedman married Rose Director, the sister of Aaron Director and an economist in her own right. She had been a fellow graduate student of Friedman's at Chicago and had also participated in the National Resources Committee's consumer study. She became and has remained an ardent collaborator in his work. He described his broadest philosophical book, *Capitalism and Freedom*, as written "with the assistance of Rose D. Friedman," and said in his preface that she had "throughout been the driving force in getting the book finished."[7] Professor Friedman appears to owe a measure of his libertarian zeal to his wife.

However, early in his career he seemed more eager to be known as a first-class economic technician and objective social scientist than as a political crusader. At the National Bureau of Economic Research, during the academic year 1939–1940, he met Harold Groves, an economics professor at the University of Wisconsin, a distinguished student of public finance, and a liberal hero to a generation of younger public-finance economists, including his students Walter W. Heller, chairman of the Council of Economic Advisers under Presidents Kennedy and Johnson, and Joseph A. Pechman, director of economic studies at the Brookings Institution.

Harold Groves was impressed with Friedman and was instrumental in arranging for him to come to the University of Wisconsin as a visiting professor for the 1940–1941 academic year. Groves clearly hoped that Friedman would stay in Madison. At the time Wisconsin was, under the still strong influence of John R. Commons and his students Selig Perlman, Edwin E. Witte, and others, the center of American institutional economics. The Wisconsin tradition was also "liberal" in the twentieth-century sense, favoring social reform, government intervention, and a moderately populist or socialist program.

The image which Friedman's arrival at Wisconsin suggests is that of a libertarian Daniel walking into the institutionalist lions' den—a historic confrontation between the "Chicago" and "Madison" doctrines. In fact, there was no such confrontation. Although Friedman's ideological sympathies lay with the free market, he had not yet actively taken up the cudgels in its behalf. His association with Wesley Mitchell—the student and admirer of Thorstein Veblen, and himself the most "scientific" of American institutionalists—gave Friedman impeccable credentials. The young Friedman had, moreover, worked primarily as a mathematical statistician. From a professional standpoint, he believed in a "value-free" economics; he was politically inert. Nevertheless, Friedman did become embroiled in controversy at Wisconsin.

One issue exercising the Wisconsin economics department

in 1941 (there was usually some issue or other to divide the department) was whether it should merge with the business school. The department was split into two camps, with the conservatives favoring the merger and the liberals opposing it. In the midst of the dispute, Friedman received and accepted the offer of a longer-term appointment; that offer then became an additional bone of contention between the two sides. Ironically, those who opposed Friedman's appointment —the forces led by Professor Walter Morton, who favored the proposed merger—were the political *conservatives* with whom Friedman was ideologically most in sympathy. The *liberal institutionalists,* the proponents of a strong and independent economics department, wholeheartedly supported Friedman. Walter W. Heller, a liberal graduate student, championed Friedman's cause and organized picket lines in his behalf.

An ugly note of anti-Semitism crept into the controversy. Selig Perlman, a distinguished labor historian, was the only Jew on the Wisconsin faculty, and some members of the department felt that one was enough. Further, Friedman was "from New York," "from the East," "from Chicago," and was thus regarded as a kind of interloper. Professor Witte, the department chairman, was indecisive. Friedman later said that Witte demonstrated "the courage of a fish." In the end, Friedman withdrew his acceptance and returned to New York.

In 1941 he completed work on his doctoral dissertation. His degree was held up, however, because of opposition from within the National Bureau of Economic Research. Most of Friedman's findings were innocuous enough; he did, however, note a substantial difference in the income levels of physicians and dentists. This difference, he argued, could be traced to the limitation in the number of physicians which occurred as the result of the establishment of elaborate requirements for medical education in the second decade of the

twentieth century. The rapid increase in training requirements, Friedman reported, "was accompanied by a marked
decline in the number of students. In consequence, the number of physicians had increased far less rapidly than total
population during the preceding 30 or 40 years, whereas the
reverse had occurred in most other professional pursuits. Indeed, the absolute number of physicians declined from 1910
until about 1920, and in 1930 was not much higher than in
1910."[8] This restriction, Friedman found, had pushed the
cost of medical care above that of dental care; and he suggested that the resulting higher return in medical services had
been intended by the medical profession.

Some members of the Bureau regarded this as an attack on
the American medical profession and demanded its removal.
Wesley Mitchell supported Friedman's analysis but was
averse to the Bureau's publishing anything on which there
was not substantial consensus. This jeopardized Friedman's
doctoral candidacy, because at that time Columbia University
required that all economics doctoral dissertations be published. Friedman was thus impelled to moderate the harshness
of his judgments and bury them deep within the lengthy
study. He also added two rather puerile alternative possibili-
ties that might explain the relative difficulty of entrance into
the medical profession: the lack of innate ability on the part
of most applicants, and a scarcity of training facilities. Only
then did Friedman advance the possibility that "the difference
in ease of entry reflects a deliberate policy of limiting the
total number of physicians to prevent so-called 'overcrowd-
ing' of the profession."[9] To be sure, he added, "an adequate
judgment of this explanation would be exceedingly difficult
and is well outside the scope of this study. . . ." Whatever the
explanation for higher medical costs, greater difficulty of
entry "has made possible or has maintained a level of income
in medicine exceeding that in dentistry by more than can be
attributed to the free working of the much-abused law of
supply and demand." Friedman was never again to be so

circumspect in his views of the impact of the medical profession or any other restrictive group on income shares or costs to consumers.

By the time a sufficient consensus on Friedman's thesis could be obtained at the Bureau, the war intervened to delay its publication until 1945, and Friedman did not get his doctorate until 1946. The episode, besides being personally obnoxious to Friedman, is significant because it marks his firm entrenchment in the free-market camp. He had discovered the effectiveness of applying simple economic theory to complex institutional phenomena. He had been impressed with the evil effects of the seemingly progressive action of regulating medical education for the stated purpose of improving medical care. And he was appalled by the resistance of even the economically sophisticated to the implications of his analysis, especially when it offended a strong interest group. The enemies of the free market were powerful and numerous; there was a war to be waged in the name of the "much-abused law of supply and demand." The application of such supply-and-demand analysis to the entire spectrum of social reality would become his main stock in trade.

Wisconsin had also left a bad taste in his mouth. His wariness of institutions, even liberal and intellectual institutions, deepened. He decided that there was more freedom and security for Jews or other minority group members in the marketplace than in "institutions"—whether they were liberal or conservative, academic or political.

I V

BACK IN New York after his year at Wisconsin, Friedman accepted an offer from Professor Carl Shoup of Columbia University to participate in a study of how to estimate future inflation and determine the amount of taxation required to prevent it. Inflation was widely anticipated in 1941 because of greatly increased military spending; Shoup had obtained a grant from the Carnegie Corporation of New York for the study. Friedman and Ruth Mack of the National Bureau of Economic Research became his associate directors.

Until then, Friedman had not been occupied with macro-economic theory, the development of models to show the workings of the national economy as a whole and its changing levels of employment, income, and prices. The new project, published as *Taxing to Prevent Inflation*, represented his first attempt to come to terms with the quantity theory of money and the new Keynesian analysis. The centuries-old quantity theory of money maintains that the level of prices is directly proportional to the quantity of money. The concepts underlying the theory were clarified by Professor Irving Fisher of Yale through his "equation of exchange": $MV = PT$, in which M is the quantity of money, V the velocity of its circulation (the number of times money turns over in a year), P the price level, and T the value of real output. The equation is a truism, since both sides describe the same phenomenon—on the left side, the amount of money spent for goods; on the right side, the amount of money received for goods. But when the assumption is made that V remains constant and the theory is stated as $M = kP$, it ceases to be a truism and means that the price level is a function of money and directly proportional to it. Thus, simply put, if the constant k is equal to 1 and the money supply is doubled, then the price level will also double.

Friedman found this formulation far too simple and inaccurate as a description of reality. Concerned with the problem of curbing wartime inflation, he turned to the "inflationary gap" method employed in earlier British economic white papers, which he described as follows:

> At a time of relatively full employment, if the government uses additional resources someone else must use less. This transfer of resources will be accomplished without a price rise if the

increase in the government's demand for the additional resources is matched by a decrease in nongovernmental demand.
Consequently, the "inflationary gap" is the "amount of the
government's expenditure against which there is no corresponding release of real resources of manpower or materials
by some other member of the community." . . . If there is such
a "gap," the government can get the real resources, but only
by bidding up prices and thereby "forcing" others to release
the resources.[10]

To prevent inflation, wrote Friedman, nongovernmental
demand would have to be reduced through taxation by an
amount equal to increased governmental demand.[11]

The first task was to decide when an inflationary gap was
in the offing. This entailed a number of hard questions,
quickly leading the analysis from the budget and the total
resources available in the society into the *monetary* area.
Friedman used a quotation from a British white paper by F.
W. Paish to indicate the range of issues involved:

> The question of exactly what constitutes an inflationary gap is
> extremely difficult to answer. Broadly speaking, it involves
> finding the answers to at least four subsidiary questions: (1)
> Has the financing of the budget deficit been accompanied by
> an increase in the quantity of money or bank deposits? (2)
> Do the people who own these increased balances wish to save
> as much, or are they intending to spend them on consumption
> goods at the first opportunity? (3) If they do wish to save as
> much, do they wish to continue to hold their savings in this
> form? (4) If they do spend them, are there sufficient resources
> available to enable the increased demand to be met without a
> serious rise in prices? In the present case the answers to the
> first and last questions are clear. There *has* been a very
> marked increase in the volume of bank deposits, and there are
> *not* the resources available to enable these increased deposits
> to be spent without a serious rise in prices.[12]

This passage is useful for understanding the later development of Friedmanian macroeconomics. The first question in-

troduces the issue of the quantity of money and points up a distinction which Friedman has always been at pains to emphasize: *the way a budget deficit is financed, and its impact on the money supply, is the heart of the inflationary matter.*

Friedman's proposal of a full-employment deficit in the federal budget (an idea advanced in his 1948 paper "A Monetary and Fiscal Framework for Economic Stability") involves such a "fiscal" expansion of the money supply to satisfy the full-employment assumption of the gap analysis. But this does not solve the issue raised by question (4): whether the impact of increasing monetary demand will be met by greater real output or by rising prices. Friedman has never adequately distinguished between the *real output* and *price* effects of monetary expansion; this is a great weakness of monetary models in forecasting short-run economic change.

Forecasts of "nominal" gross national product—total output in current dollars—are of slight value to businessmen or government policymakers if one cannot say how much of the anticipated rise in total output and income will be real and how much will merely represent higher prices. Friedman has been ambiguous concerning the short-run forecasting potential of money-supply changes. For the longer run, he is not ambiguous at all; money-supply changes, he says, uniquely account for long-term price changes. His conclusion for the 1943 study, *Taxing to Prevent Inflation,* expresses this curious mixture of caution and confidence:

> This "inflationary gap" analysis does not give any specific technique for determining the rise in income or prices that will be associated with a given increase in government expenditures. At the same time, it seems exceedingly valuable. It furnishes a significant framework for interpreting government policy, and it reveals the basic factors on which its effects depend. It combines the monetary and "real" aspects of the problem in a way that gives each its proper place.[13]

But in 1941, Friedman was by no means giving primacy to monetary factors. In a critical response to what he regarded as an overly optimistic appraisal of inflationary gap analysis by Walter Salant, Friedman largely ignored monetary factors as a cause of inflation.[14] Ten years later, when Friedman brought out his *Essays in Positive Economics,* he felt compelled to remedy this omission:

> The next seven paragraphs and the subsequent material enclosed in brackets are additions to the article as originally published. As I trust the new material makes clear, the omission from that version of monetary effects is a serious error which is not excused but may perhaps be explained by the prevailing Keynesian temper of the times.[15]

There is no doubt that, during the war years, while he was working with Professor Shoup of Columbia, Friedman was influenced by what he calls "the Keynesian temper of the times." The British economist Joan Robinson has pointedly suggested that Friedman, together with his monetarist disciples, are among the "bastard progeny" of Keynes—because, despite their emphasis on monetary rather than fiscal factors, they too believe that regulating aggregate demand will cure the basic ills of the capitalist system. In any case, Friedman was certainly no "monetarist" when he served in Washington during World War II.

In the fall of 1941, Friedman had followed Shoup to the Treasury, where he became principal economist of the Division of Tax Research. There he was involved in drafting the major tax reform legislation of the early 1940s. As a by-product, he acquired a certain contempt for the legislative processes of the national government. As he later recalled:

> In the course of a presentation to the House Ways and Means Committee on the need for additional taxes to prevent inflation, I was interrupted by one member who exclaimed, "Why do we need to worry about inflation in considering taxes? We

have just passed General Max [the measure that put 'general maximum' ceilings on all wages and prices]. It is now up to Leon Henderson [director of the Office of Price Administration] to control inflation." I had barely embarked on a learned discourse about how General Max would not work unless it was reinforced by measures to reduce purchasing power, when he interrupted me again. "I understand that," he said. "Mr. Henderson may fail, but we have discharged our responsibility by giving him the power. Now it's up to him."[16]

Here Friedman was in agreement with Leon Henderson's chief lieutenant at OPA, John Kenneth Galbraith. But, although Galbraith was and has remained a strong supporter of rationing and controls—provided they are not the sole weapons used to prevent inflation—Friedman was personally against controls.

Friedman favored the "spendings tax" proposed by Allen Wallis, his friend and fellow graduate student at Chicago. This was a tax not on income but on expenditure; it would have the advantage of encouraging saving and reducing purchasing power without requiring the bureaucracy needed to enforce controls. The tax would be collected on the fraction of total income a consumer spent, rather than on his purchases of particular items, as with a sales tax. Its superiority to a sales tax was that it could be made progressive. One of its drawbacks was the need for it to be collected at least quarterly if it were to be immediately effective against inflation; another was the difficulty of ensuring the accuracy of taxpayers' statements. For these reasons, Carl Shoup had rejected the spendings tax in *Taxing to Prevent Inflation*.

Friedman nonetheless found the spendings tax appealing because, while holding down inflation, it only minimally disturbed the operation of the market. In 1943 he wrote a paper recommending adoption of the spendings tax as part of an integrated fiscal program adapted to wartime needs, rather than as an alternative to direct controls.[17] Whatever his feelings about controls at the time—he seems to have accepted them as a wartime inevitability, as did a young lawyer in OPA, Richard M. Nixon,—in retrospect, Friedman, like Nixon, condemned them harshly:

> During World War II, when price control was nearly universal, black markets, and rationing by chance, favoritism, and bribery developed in steel, meat, bananas—you name it.[18]

> Even 60,000 bureaucrats backed by 300,000 volunteers plus widespread patriotism were unable during World War II to cope with the ingenuity of millions of people in finding ways to get around price and wage controls that conflicted with their individual sense of justice.[19]

By 1943, work on tax reform had for the most part been completed, and Friedman decided that he could best help the war effort by joining the Division of War Research's statistical research group, headed by Allen Wallis, then at Columbia. Memories of college ROTC left him with no desire actually to join the Army. Friedman's work at Columbia bore no relation at all to economics; rather, it consisted of statistical analysis for such projects as the development of high-temperature alloys and the proximity fuse. By the end of the war, Friedman's skill as a mathematical statistician reached a never-to-be-regained peak.

V

AFTER THE WAR, Friedman spent a year as associate professor of economics at the University of Minnesota, then joined the faculty at Chicago, where he has remained. The Chicago economics department had changed substantially in the decade since Friedman's student days. Paul Douglas, who returned to academic life there after a distinguished career in the Marine Corps, has described this change well in his autobiography:

> I was disconcerted to find that the economic and political conservatives had acquired an almost complete dominance over my department and taught that market decisions were always right and profit values the supreme ones. The doctrine of noninterference with the market meant, in practice, clear the track for big business. Inequalities of bargaining power, knowledge, and income were brushed aside, and the realities of monopoly, quasi monopoly, and imperfect competition were treated as either immaterial or nonexistent. Similarly, conflicts of interest between producers and consumers were also brushed aside, as were the possibilities of private firms unloading social costs upon the community and the environment. Polluted air and water and excessive noise seemed unimportant to the dominant conservatives. Furthermore, since market demand was based on the distribution of income, and really of surplus income above the minimum of subsistence, it reflected all the injustices of modern society and the thinking of Ruskin and Hobson.
>
> I could not accept this as the complete gospel. The dislike of government, which was sound in the case of permanent price-fixing, was carried over into many fields where the state seemed to be a good agency for widely needed reforms, in health, housing, education, conservation, and recreation. The opinions of my colleagues would have confined government to the eighteenth-century functions of justice, police, and arms, which I thought had been insufficient even for that time and were certainly so for ours. These men would neither use statistical data to develop economic theory nor accept critical analysis of the economic system. Though expounded with intellectual subtlety, their unrealistic view did not furnish adequate answers to the problems that beset us. It was too much like the economics of the period prior to World War I. So I found myself increasingly out of tune with many of my faculty colleagues and was keenly aware of their impatience

and disgust with me. The university I had loved so much seemed to be a different place. Schultz was dead, Viner was gone, Knight was now openly hostile, and his disciples seemed to be everywhere.[20]

Henry Simons, the vigorous opponent of monopoly, was dead, and Knight had indeed become cranky and idiosyncratic. Soon Douglas himself was gone: in 1948 the voters of Illinois elected him to the United States Senate. Friedman was, of course, part of the new wave which Douglas deplored.

Upon his return to Chicago, Friedman came out as an unabashed crusader for the free market. Together with George Stigler, he had written *Roofs or Ceilings? The Current Housing Problem,* a tract against postwar rent control. In it, Friedman and Stigler argued cogently that the free market would provide the best solution to the acute housing shortage left in the wake of World War II:

1. In a free market, there is always some housing immediately available for rent—at all rent levels.
2. The bidding up of rents forces some people to economize on space. *Until there is sufficient new construction, this doubling up is the only solution.*
3. The high rents act as a strong stimulus to new construction.
4. No complex, expensive, and expansive machinery is necessary. The rationing is conducted quietly and impersonally through the price system.[21]

Roofs or Ceilings? was brought out under the auspices of Leonard Read's conservative Foundation for Economic Education. Friedman and Stigler soon became aware of the hazards of association with free-market ideologues outside the economics profession. Other economists attacked them for linking their names to what was considered a blatantly propagandistic organization. At the same time, however, the sponsors of the Foundation for Economic Education were suspicious of what they saw as hints of unorthodoxy on the part of the authors. For Friedman and Stigler had shown untoward sensitivity to the charge that the free market favored the rich, and they had even gone so far as to support greater economic equality:

> The fact that, under free market conditions, better quarters go to those who have larger incomes or more wealth is, if anything, simply a reason for taking long-term measures to reduce the inequality of income and wealth. For those, like us, who would like even more equality than there is at present, not alone for housing but for all products, it is surely better to attack directly existing inequalities in income and wealth at their source than to ration each of the hundreds of commodities and services that compose our standard of living. It is the height of folly to permit individuals to receive unequal money incomes and then to take elaborate and costly measures to prevent them from using their incomes.[22]

Such egalitarian sentiments could not be allowed to stand alone, and an "editor's note" was appended by the Foundation for ideological clarification:

> The authors fail to state whether the "long-term measures" which they would adopt go beyond elimination of special privilege, such as monopoly now protected by government. In any case, however, the significance of their argument at this point deserves special notice. It means that, even from the standpoint of those who put equality above justice and liberty, rent controls are the height of folly.[23]

The note was probably right in its assessment of the argument's significance; Friedman and Stigler were nonetheless infuriated at its inclusion. For years they had nothing more to do with Leonard Read and his Foundation. For Friedman, however, there would remain the worry of how to treat his more extreme bedfellows in the emerging libertarian movement.

The criticism which *Roofs or Ceilings?* provoked among other economists was another matter. The pamphlet, as Paul Samuelson later observed, "actually outraged the profession —that shows you where we were in our mentality in the immediate postwar period." The profession as a whole had drifted very far from an adequate appreciation of the role of the market and of prices as a means of efficiently allocating scarce resources. In their outrage, many economists seized upon the "interest group" sponsorship of the Friedman-Stigler pamphlet as evidence of its "corruption." The authors had been paid, as they acknowledged, for writing this libertarian tract!

The outcry over *Roofs or Ceilings?* raised for Friedman the difficult issue of how to resolve the competing impulses toward public advocacy and toward the kind of scientific objectivity and scholarly consensus so fervently espoused by Wesley Mitchell. It also raised the deeper issue of value conflicts among economists: Were ethical values extraneous to or inherent in economic analysis itself?

Distressed with his profession and with what he regarded as confusion over the nature of economics as a science, Friedman wrote "The Methodology of Positive Economics," an essay which became the first in a collection daringly titled *Essays in Positive Economics*.[24] Drawing on the distinction of John Neville Keynes (John Maynard Keynes's father) between a *positive science*, a *normative science*, and an *art*, Friedman asserted:

> Positive economics is in principle independent of any particular ethical position or normative judgments. As Keynes says, it deals with "what is," not with "what ought to be." Its task is to provide a system of generalizations that can be used to make correct predictions about the consequences of any change in circumstances. Its performance is to be judged by the precision, scope, and conformity with experience of the predictions it yields. In short, positive economics is, or can be,

an "objective" science, in precisely the same sense as any of the physical sciences.[25]

As positive economics progressed, Friedman suggested, normative differences (that is, differences over economic policy) would tend to disappear, or at least narrow. For positive economics would yield policy conclusions "that are, and deserve to be, widely accepted."[26] Friedman's strong implication was that economists who disagreed with him were guilty not only of questionable values or political purposes but also of an unscientific method, a neglect of "positive economics."

But what, precisely, was the nature of this positive economics? It was the formulation of hypotheses capable of yielding predictions that could be tested; the correctness of the prediction established the fundamental validity of the hypothesis. Such predictions could, of course, be retrospective—that is, they did not have to be forecasts of future events but could be predictions "about phenomena that have occurred but observations on which have not yet been made or are not known to the person making the prediction."[27]

Friedman believed that, in stressing the construction of hypotheses with predictive value as the main business of economics, he was following in the tradition laid down by the great British economist Alfred Marshall. He contrasted this Marshallian approach with that of Leon Walras, the Swiss economist who had drawn up a logically consistent system of equations, devoid of empirical content, to describe general economic equilibrium, the state of rest to which the system was tending. Friedman viewed "pure" economic theorists as Walrasian; as a rule, he found their work vacuous. But he also regarded those economists who set up elaborate systems of equations employing real data (the evolving science that bore the name "econometrics") as Walrasian; they, too, were mere builders of fancy models that lacked positive, predictive value. He cast Wassily Leontief, with his elaborate input-output matrices of the economic system, into this Walrasian Hell.

Predictive value was, however, only part of the story; the most controversial part of Friedman's concept of positive economics was still to come. For it was not merely that the validity of hypotheses *could* be tested by the accuracy of their predictions; rather, this was the *only* way they could be evaluated. To judge a hypothesis by the realism of its assumptions would be illicit. Thus, he contended, it is wrong to reject the hypothesis of free-market price determination in an industry if it turns out on inspection that the industry is not perfectly competitive; what matters is whether prices are set *as if* there were perfect competition. Friedman went so far as to make a positive virtue out of "unrealistic" assumptions:

> In so far as a theory can be said to have "assumptions" at all, and in so far as their "realism" can be judged independently of the validity of predictions, the relation between the significance of a theory and the "realism" of the "assumptions" is almost the opposite of that suggested by the view under criticism. Truly important and significant hypotheses will be found to have "assumptions" that are wildly inaccurate descriptive representations of reality, and, in general, the more significant the theory, the more unrealistic the assumptions (in this sense). The reason is simple. A hypothesis is important if it "explains" much by little, that is, if it abstracts the common and crucial elements from the mass of complex and detailed circumstances surrounding the phenomena to be explained and permits valid predictions on the basis of them alone. To be important, therefore, a hypothesis must be descriptively false in its assumptions; it takes account of, and accounts for, none of the many other attendant circumstances, since its very success shows them to be irrelevant for the phenomenon to be explained.[28]

He claimed that assumptions could be used only to specify the circumstances under which a theory would hold, not to determine them. That could be accomplished only by a test of predictive accuracy. Here Friedman was aiming in a different direction—at those economists, such as Leontief, who sought to improve economic theory by making it more "realistic." They were guilty, he felt, of confusing descriptive accuracy with analytic relevance. Prime representatives of this group were those who complicated and confused the greatest discovery of economics, the functioning of markets. And the principal villains were E. H. Chamberlin of Harvard and Joan Robinson of Cambridge University, whose theories of monopolistic or imperfect competition had been widely praised and accepted by the economics profession for bringing economic theory in closer touch with reality.[29] Friedman, Frank Knight's ally and Paul Samuelson's antagonist, held that, however much Chamberlin's or Robinson's models might seem more realistic than the traditional model of economic competition, they were actually of little analytical value:

> The theory of monopolistic competition offers no tools for the analysis of an industry and so no stopping place between the firm at one extreme and general equilibrium at the other. It is therefore incompetent to contribute to the analysis of a host of important problems: the one extreme is too narrow to be of great interest; the other, too broad to permit meaningful generalizations.[30]

Friedman's methodological insistence on the free market as the basic hypothesis of positive economic analysis conveniently dovetailed with his ideological faith in laissez-faire and his opposition to government controls and planning.

At the end of his essay, Friedman summed up the current state of economic theory. Relative price theory was in good shape. So was static monetary theory; based on the quantity theory of money, it was designed to explain the price level and the dollar value of the economy's total output. Least satisfactory was monetary dynamics, the process by which the economy as a whole adapted to changes in monetary conditions.[31]

On Friedman's own estimate of what economics had accomplished, his methodology of positive economics emerges as backward looking or even reactionary. The simple, predictive hypotheses which Friedman advocated could be adequate only in those areas where economics had already been successful—relative price theory and monetary statistics. The likelihood was slight, however, that simple hypotheses could be found to explain complex processes of change in the national or world economy. And where the free market was lacking and monopolies dominated, the "unrealistic" hypotheses of competitive markets simply did not lead to correct predictions. For instance, in the summer of 1974, Friedman predicted that the international oil cartel would soon break up and prices would fall close to where they were before the Yom Kippur War.[32] Two years later, it still had not happened.

Simple economic models, rules of thumb, blueprints of the future, "laws" based on highly restrictive cases from the past —all were inadequate to deal with complex and changing reality. Friedman's positive economics fails in its own terms; his simple and positive predictions are often wrong. An economic model—being only a simile, as Georgescu-Roegen has said—cannot serve as a substitute for well-founded hypotheses, detailed knowledge of the actual situation being analyzed, and that most uncommon of all assets to the analyst, common sense and sensitivity. "An artless analysis cannot subserve an art."[33]

VI

FRIEDMAN is most famous for his work on money and his discovery of the doctrine that came to be called *monetarism*. This doctrine has two basic tenets: (1) change in the money supply is the only systematic factor influencing the overall level of spending and economic activity; hence (2) the only action required to ensure prosperity and price stability is for the central bank to stabilize the rate of growth of the money supply at a rate approximately equal to the real rate of growth of the economy—say 4 to 5 percent per annum. The theory effectively excludes changes in tax rates, in government spending, or even in private investment and consumption as systematic influences on the economy.

The underlying motivation for Friedman's discovery of monetarism appears to have been his general dislike of Keynesianism and his aversion to any economics which, by sanctioning controls or otherwise opening the door to government intrusion in the free market (and thus to socialism), might undercut the intellectual preeminence of individualism and freedom of enterprise. But the proximate cause of his discovery of monetarism was simply a parceling out of work at the National Bureau of Economic Research, with which Friedman continued to be associated after his return to the University of Chicago.

It had become clear that the aging Wesley Mitchell would never be able to complete his last extensive study of American business cycles. Something, however, had to be done with the great mass of material which he had compiled for it. Arthur Burns, Mitchell's successor as research director at the Bureau, assigned aspects of the study to various members of the research staff; Friedman drew American monetary history.

As he began to work on what eventually became *A Mone-*

tary History of the United States,[34] his belief in the importance of monetary theory and policy grew. This can be traced through the papers reprinted in his *Essays in Positive Economics*. In addition to revamping the inflationary-gap article, he offered an extended discussion of proposals for a commodity-reserve currency. Henry Simons's proposal of 100 percent reserve banking is incorporated into "A Monetary and Fiscal Framework for Economic Stability." Friedman's plan, which he hoped would provide "a minimum program for which economists of the less extreme shades of opinion can make common cause," had four elements: (1) a reform of the monetary and banking system to eliminate both the private creation and the private destruction of money (that is, the establishment of 100 percent reserve banking), with discretionary control over the quantity of money vested in the central bank; (2) a policy of determining the volume of government expenditures on goods and services entirely on the basis of the community's desire, need, and willingness to pay for public services; (3) a predetermined program of transfer expenditures by government, stating the conditions and terms under which relief, assistance, and other transfer payments would be made; and (4) a progressive tax system that would put primary reliance on the personal income tax.[35]

In this 1948 paper, Friedman made clear his opposition to discretionary action—fiscal or monetary—to offset cyclical change. His one concession to the Keynesians was an early version of the full-employment budget; he would gear government spending to what tax revenues would be when the economy was operating at a high level of employment, and would accept deficits when the economy was operating below that level, or surpluses in times of boom and overly full employment. When, early in 1971, President Nixon disclosed his intention to propose an unbalanced budget to lift the economy back to full employment, his economic advisers could point to Friedman's early statement of the full-employment, balanced-budget principle as evidence for the doctrine's conservative lineage.

In "Comments on Monetary Policy" (1951), Friedman again addressed the problem of inflation, newly troublesome as a result of the Korean war. He advocated the use of both fiscal and monetary policy to prevent inflation. Only in his recommendation that the effect on the rate of interest be ignored by monetary policymakers did he anticipate his coming position that the Federal Reserve should ignore interest rates and concentrate solely on the rate of growth of the money supply. Friedman was on his way back to the quantity theory of money. As he moved to rehabilitate monetary policy as the prime means of stabilizing the economy, Friedman suggested that the quantity theory, properly understood, was equivalent to Keynesian theory; monetary policy to frustrate spending and thereby prevent inflation, he said, could be described in either of two alternative languages—that of the quantity theory or that of Keynesian analysis.

But that was in 1951, and as the 1950s wore on, Friedman moved to use his new monetarism as the basis for a Chicago School counterattack on the Keynesian revolution—a would-be "counter-revolution," as Professor Harry Johnson of the University of Chicago has put it. In 1956, Friedman made a wholesale endorsement of a modern, sophisticated version of the quantity theory of money; his theory, he contended, was the outgrowth of a unique tradition of sophisticated and "modern" quantity theorizing practiced by his former teachers at Chicago. Don Patinkin, who was a fellow graduate student of Friedman's at Chicago and a specialist in the monetary area, has neatly disposed of this contention.[36] In conversation, Friedman has conceded that Patinkin was correct, but only so far as the theory itself was concerned. What was really important, Friedman claimed, was the broad emphasis on monetary factors at Chicago; this insulated the students from the Keynesian revolution and its downgrading of monetary factors. The claim seems a bit hollow in the light of Friedman's own neglect of monetary factors during the war years, when, as he said, he was influenced by "the Keynesian spirit of the times."

Nevertheless, Friedman is right to stress the importance of the Chicago tradition—typified by Frank Knight—as the key factor in his opposition to the Keynesian revolution. But what really spurred Friedman and other Chicagoans to attack the Keynesians was not the technical aspects of monetary theory but the Chicago tradition of political conservatism and the ideology of laissez-faire.

The real "old-time religion" of the economic conservatives was that the government should stay clear of the economic process, including the creation of additional money. For, the argument went, any change in the quantity of money would cause a distortion in the economy, which could be put right only by a reverse change in the quantity of money. But government could not be depended upon to meddle with money. Given their spending proclivities, kings and politicians, they would always create too much money, breed inflation, destroy the frugal habits of good citizens and the prudent investments of the responsible, and create conditions that would force government to play a larger and larger role in the system. In due course, inflation would lead to economic collapse, climbing unemployment, and hence an even larger role for the state. The best thing was simply to keep the government's hands off the money printing press—to let true money come out of gold mines, and to let credit be created by free citizens doing business among themselves.

This was the doctrine that Abba Lerner has called "100 percent pure capitalism."[37] But to Friedman it had one serious difficulty. If the economy was to expand while the stock of money remained basically unchanged, prices would have to fall continuously; in a modern society, with strong resistances to price and wage reductions, it might be too much to expect that deflation could be made consistent with high employment and steady economic growth. On the contrary—on the basis of the experience of the Great Depression—deflation (falling wages and prices) seemed far more likely to lead to falling output and rising unemployment.

Hence, Friedman's proposal that the money supply be held to a constant rate of increase of 4 or 5 percent a year represented an abandonment, but a minimal abandonment, of the pure capitalist position. It yielded to the difficulty (or impossibility) of enforcing wage and price deflation. But, in establishing a *rule* for monetary growth, it would tie the hands of government and the monetary authorities, prevent inflation by permitting just enough increase in money and spending each year to cover normal increases in output, and thereby reduce or even eliminate the need for government to take specific actions to deal with unemployment or inflation.

Thus, laissez-faire could be reconstituted in modern form. The Keynesian revolution's imposition upon government of the "socialistic" responsibility of maintaining full employment would be overthrown, together with its inflationary threat to capitalist society; and so would the New Deal's detailed meddling in particular markets for goods or services, done in the name of restoring or preserving prosperity and high employment. None of this was necessary any longer; steady monetary growth, federal deposit insurance, and a few built-in stabilizers (including a full-employment, budget-balancing rule) would ensure against another depression, and the market system was best equipped to handle almost everything else.

To be sure, there were problems. What if there actually were inflation, for whatever reason—excess demand resulting from past policy mistakes, inflationary expectations, the hangover of past inflations, the stimulus of military buildups, the power of labor unions or international cartels, or widespread droughts or other shortages? Would the fixed money supply rule work? If the supply of money were permitted to

Milton Friedman

increase by, say, only 5 percent a year and prices rose by 10 percent a year, the real volume of spending measured in purchasing power would drop—and so would real output and employment.

The monetarists had a simple answer to this. Inflation resulted from one cause alone: too much money. Extraneous factors like wars, droughts, or strikes could not cause inflation, unless they prompted the monetary authorities to increase the money supply too fast. If the authorities refused to do that, there might be short-run pains of adjustment, but sooner or later the economy would get back on the track of steady, noninflationary growth. Such were the recuperative powers of a free economy with a flexible price system.

There were, however, differences among monetarists on whether to use monetary policy to offset other short-run factors making for economic instability. The "real cash balance" school favored fine tuning—for instance, to lift the economy out of a slump or to check a rapid inflation. But the more conservative and orthodox monetarists (led by Friedman himself) concluded that such discretionary changes would do more harm than good, contending that efforts to "fine tune" monetary policy would themselves become a source of instability and disturbance. Friedman stated his own simple and restrictive rule for monetary policy in his *Capitalism and Freedom* (1962):

> I would specify that the Reserve System shall see to it that the total stock of money . . . rises month by month, and indeed, so far as possible, day by day, at an annual rate of X per cent, where X is some number between 3 and 5. The precise definition of money adopted, or the precise rate of growth chosen makes far less difference than the definite choice of a particular definition and a particular rate of growth.[38]

Friedman's professed ignorance about the precise workings of the business cycle (which has not always inhibited him from making very definite cyclical predictions on the basis of money-supply changes) constitutes the crucial connection between his monetarism and his belief in laissez-faire. Government stabilization policy was almost sure to be wrongheaded; he alleged that the Federal Reserve's contraction of the money supply had brought on the Great Depression. And when the monetary authorities acted on the basis of their analyses of business conditions, they came on too late and too strong—whether to reverse a slump or to check a boom. "Steady as you go" became the watchword of true monetarism.

This monetarist philosophy obviated the need, in the view of its proponents, for government to deal dramatically or urgently with institutional distortions resulting from labor unions, corporate monopolies, or other groups with market power. True, such distortions created problems, but economic stability and prosperity could be achieved nonetheless, since the market was the dominant force in society, not the interferences or distortions resulting from monopoly elements. Here Friedman had to tread a delicate course between accepting monopoly forces and criticizing them. For if he implied too strongly that there was a serious interference with free-market behavior at the present time, then he would call into question the applicability of libertarian monetarism. Yet he could not be too light on monopoly power, since his basic aim was to restore the free market. Because of the contradictions involved, Friedman was often reduced to statements like this:

> Labor unions are important political and economic institutions that significantly affect both public and private actions. This fact raises serious and difficult problems for economic policy. At the same time, laymen and economists alike tend, in my view, to exaggerate greatly the extent to which labor unions affect the structure and level of wage rates. This fact is one of the most serious obstacles to a balanced judgment about appropriate public policies toward unions.[39]

In other words, market power in the hands of unions was undesirable, but efforts of government to control that power (and regulate wages) would be worse—a still more inefficient distortion of market-determined wages and a swelling of truly dangerous power in the hands of the government. Friedman felt exactly the same way about government efforts to curb corporate power, a position that increased his popularity with big businessmen and anti-antitrusters.

Friedman did not think it possible, nor especially desirable, for economics to provide a comprehensive, detailed description of an economy; it was as though he feared that such data would lead to an abridgement of human or business freedom. In any case, the knowledge was unobtainable; economic data, desires, and decisions were subjective, in the minds of individuals. What society needed were positive solutions to (hitherto) normative questions of a relatively simple, basic, and overriding character. The object was to substitute "correct" long-term policies for difficult and unpleasant short-run decision making. That was what monetarism was all about; instead of permitting some government bureaucrats to determine the suitable balance between rates of inflation and unemployment, a gradual growth in the money supply should be required. That was the answer to economic stability. Any resulting hardships (such as unemployment or unequal distribution of income) could be alleviated—if society so desired—by other means.

VII

AS the postwar years wore on, Friedman assumed Frank Knight's mantle as head of the Chicago School of economics, although as a national figure Friedman won far greater fame. With the sword of laissez-faire, he sought to cut knot after

Gordian knot of economic problems. In the international monetary area, he championed freely floating exchange rates long before the rest of the economics profession awoke to the dangers and imminent breakdown of a system of fixed exchange rates. The collapse of the postwar Bretton Woods international monetary system did not lead to a collapse of the world economy (as had happened in 1931) thanks to the intellectual groundwork that had been laid, in large measure by Friedman, for acceptance of floating rates.

He opposed government subsidies and government regulation of industry, and helped to open the eyes of his own profession—and then of politicians—to the wastefulness and inefficiency of much regulation, especially when it propped up price fixing and anticonsumer and anticompetitive behavior by business. He applied his principles of free-market pricing and voluntarism to many important new areas. For example, he argued for and eventually won the case for an all-volunteer army. With many young people in bitter protest against the Vietnam war and the draft, the volunteer army helped quiet the campuses and ease a burning national problem. At first, it appeared that the concept was turning out to be a very costly failure; Friedman himself blamed the military for putting its worst officers into the personnel recruitment jobs. But the serious slump and high unemployment of 1974–1975 turned the idea into an apparent success, as the quality, discipline, and morale of the armed services improved.

He did not have an unbroken string of victories, however. Friedman argued hard for an educational voucher system that would have permitted students to attend any public school they and their family wished; in part, this was meant to be a solution to the school integration conflict. Friedman proposed the system not only as a means to get around busing but also to provide incentives for school administrators and teachers to behave more efficiently and creatively in order to attract more and better students (and funds) and to avoid losing students, public support, and their reputations. But the complexities of city life, the problems of coping with students of varying social backgrounds, and the resistances of the teachers and of community groups have thus far barred an urban test of such a voucher system.

He also proposed and campaigned hard for a "negative income tax"—a sliding scale of benefits to persons willing to work, even at very low-paying jobs—as an alternative to welfare payments and other social benefits, such as food stamps or public housing. He helped to convert many liberals as well as conservatives to the logic of providing a basic floor under everyone's income, with higher incomes to be achieved (without dollar-for-dollar losses) by those willing and able to supplement their benefits with earnings up to a certain cutoff point. The Nixon administration accepted the principle in its Family Assistance Program. Nevertheless, the legislation failed to pass, largely because of faint support from the administration and because of opposition from conservatives opposed to universal benefits and from liberals who disliked the plan's low benefit levels and its elimination of other "categorical aid" programs.

Friedman repeatedly expressed his opposition to tax breaks for industries, bailouts of failing corporations and banks, Social Security, foreign economic aid, and other government programs. His general principle was that anything government does can, with certain exceptions like maintaining law and order, managing the national defense, and overseeing currency, be done better by competitive capitalism. Even in areas where government had a role to play, it should, wherever possible, be subjected to market tests and antibureaucratic limits. Tax rates should be steadily reduced as the economy expanded, as a means of checking the growth of government. Steady, gradual increase of the money supply would have the side benefit of preventing government from reaping the larger share of national resources that would accrue from the impact of a progressive tax system on an inflation-prone economy. Indeed, he proposed that the progressive income tax (which he earlier had favored) should be replaced by one with a constant rate at all incomes, with all deductions and exclusions eliminated. This, he argued, would yield as much income as a loophole-riddled progressive income tax, and would increase incentives for saving, investment, and wise use of resources. It would also eliminate the need for an army of tax lawyers and tax accountants, and the opportunities for favoritism by government officials.

Friedman thus has emerged as one of the leading lights of the small but growing and highly vocal libertarian movement. Among libertarians, however, Friedman is not considered the fanatical devotee of laissez-faire that his fellow economists or the general public have often thought him to be; rather, he is correctly perceived by libertarians as a "limited-government man." Although he thinks well of the philosopher and novelist Ayn Rand (her novels, he said, were "very influential" in making people take seriously the ideas he considered right), he finds that "elements of intolerance and doctrinaire faith make some of her disciples impossible—intolerable!" (Alan Greenspan, chairman of President Ford's Council of Economic Advisers, is a professed Randian, albeit a relatively tolerant and undoctrinaire disciple.) Likewise, Friedman regards the late Professor Ludwig von Mises, an ultraconservative economist, as a "great man," but finds his disciples "impossible." Not merely are such extreme libertarians irresponsible, in his view, but they are living in a revolutionary fantasy. Friedman prides himself on being a realist; he thinks his talent lies in espousing creative and feasible social reforms rather than simply envisioning a capitalist nirvana.

In the backlash to the social upheavals of the 1960s, and to the New Deal, Fair Deal, New Frontier, and Great Society programs of Democratic administrations, Friedman won public prominence and became a major intellectual influence on the Republican Party and on Republican presidents and presidential hopefuls. He also made an increasingly deep impression on many members of the economics profession, especially those who were restive under the dominance of the now-orthodox Keynesian establishment.

VIII

FRIEDMAN'S most important scholarly book, *A Monetary History of the United States,* written in collaboration with Anna J. Schwartz for the National Bureau of Economic Research, appeared in 1963 and provided the underpinnings for the monetarist doctrines he had been selling in lectures, public speeches, and articles for years. This massive work received high praise from most professional reviewers, although many of them disagreed with some of its theory as well as its facts—particularly the authors' contention that the Great Depression had been caused by the Federal Reserve's reduction of the money supply. The critics noted that from 1929 to 1932 the money supply fell by less than 3 percent, but prices dropped by more than 30 percent, real investment, consumption, and output plummeted, and unemployment climbed to 25 percent of the labor force. It was difficult to see how such a catastrophe could have resulted from so small a decline in the stock of money. Despite such criticisms, the book helped swell the growing ranks of the monetarists.

In 1964, Friedman participated actively in a presidential campaign for the first time. He allied himself with Senator Barry Goldwater in March, and after the Republican National Convention, when conservative economists like Arthur F. Burns, Paul McCracken, and Henry Wallich were abandoning the GOP candidate, Friedman emerged as Goldwater's most prominent economic adviser. Unlike his former student, Professor G. Warren Nutter of the University of Virginia, who worked full-time on the campaign, Friedman did not hit the campaign trail. He wrote position papers and from time to time communicated directly with Goldwater. He also publicly expounded Goldwater's economic position. This position differed in some significant respects from Friedman's own; Goldwater, with his nationalistic and probusiness sympathies, did not agree with Friedman's support of free trade and floating exchange rates. But these differences were glossed over by Friedman in his public pronouncements; he immediately recognized that if he was going to play the political game, he would have to soften or blur his own positions when they were in conflict with his candidate's. Adviser and candidate did agree in opposing increased government spending and countercyclical fiscal policy, in advocating a reliance on monetary policy to buoy up the economy (and tax reductions to reduce the role of government), in supporting federal-state revenue sharing and a volunteer army, and in extolling the glories of capitalism and individualism.

In retrospect, Friedman now feels that Goldwater's was an "influential campaign" but that Goldwater surfaced too early as a presidential candidate. "If Nixon had not decided to run for governor in California," he says, "he would have got the presidential nomination and been defeated as a presidential candidate in 1964." Goldwater, he thinks, would then have been elected in 1968 by a greater margin than Nixon was; he thinks the issues in 1968 were "all Goldwater issues—the volunteer army, law and order, fiscal responsibility, and busing."

After the debacle of the 1964 Goldwater candidacy was over, Friedman was happy to return to academe full-time and to resume his familiar and uninhibited role of shouting warnings and encouragement from the sidelines. When *Newsweek* magazine offered him a slot as one of its regular economic commentators, he readily accepted. But when Richard Nixon became the Republican candidate in 1968, Friedman was more than willing to return to the political wars.

After Nixon's narrow victory in 1968, Friedman did not seek any office in Washington but preferred to be an outside adviser from the University of Chicago and from his country place in Ely, Vermont. His ideas played an important role— off and on—in the Nixon administration. President Nixon's first chairman of the Council of Economic Advisers, Paul McCracken, described himself as "Friedmanesque," and the basic "gradualist" policy that the administration pursued in 1969 and 1970 was designed to sweat inflation out of the system by a slow and steady rate of monetary growth that would keep the economy operating below its potential. And the Nixon administration also put through tax cuts, despite the persistence of inflation, in part to reduce government's role in the economy.

This Friedmanesque medicine did not work as well as the doctor himself expected it to. The economy sank into recession, unemployment rose, but inflation scarcely receded. With the Republican losses in the 1970 congressional elections behind him, and the 1972 presidential election ahead of him, Nixon made a dramatic switch to his New Economic Policy on August 15, 1971, imposing wage and price controls (an act that was anathema to Friedman and his fellow Chicagoans, including George Shultz, then director of the Office of Management and Budget, and Herbert Stein, the new chairman of the Council of Economic Advisers), calling for a strong fiscal stimulus (also an anti-Friedmanesque stand), but cutting the dollar loose from gold and letting the dollar float on the exchange markets (a victory for Friedman's line). At the University of Chicago, Friedman expressed his unhappiness over price controls, but he was far less bitter toward President Nixon than he had been toward those Democrats and liberals who advocated controls before the administration adopted them. In an article in the *New York Times* on October 28, 1971, he rose to high philosophical and moral ground in attacking the controls:

> The controls are deeply and inherently immoral. By substituting the rule of men for the rule of law and for voluntary cooperation in the marketplace, the controls threaten the very foundations of a free society. By encouraging men to spy and report on one another, by making it in the private interest of large numbers of citizens to evade the controls, and by making actions illegal that are in the public interest, the controls undermine individual morality.
>
> One of the proudest achievements of Western civilization was the substitution of the rule of law for the rule of men. The ideal is that government restrictions on our behavior shall take the form of impersonal rules, applicable to all alike, and interpreted and adjudicated by an independent judiciary rather than of specific orders by a government official to named individuals.[40]

These principles would appear to have applied a fortiori to the political spying, wiretapping, investigations of the tax returns of political opponents for the purpose of retribution or harassment, the attempted corruption of the CIA, FBI, Internal Revenue Service, and other government agencies—the bundle of offenses known as "Watergate" that toppled the Nixon administration. But here Friedman and many other libertarians were far more restrained. Privately they expressed their unhappiness to each other, but publicly they were mild in criticizing the president and his aides, who had so traduced the principle of the "rule of law" and had substituted the "rule of men."

In the narrower area of economic policy, Friedman did concede, in an address before his colleagues at the American Economic Association on December 28, 1971, that he had been "punished" by the experience of watching his prescribed monetary policy fall short of his boldly advertised expectations.[41] Given the changes in monetary policy adopted by the Federal Reserve in 1969, 1970, and much of 1971, said Friedman, the changes in "nominal" national income—expressed in current dollars—had been about what he and other monetarists had expected. But, he admitted, he had not expected so much of the decline in "nominal" income to be in the output of real goods and services, and so little in the rate of inflation. "On this issue," he said, "I must confess that I made overly optimistic predictions in 1969 about how soon inflation could be expected to respond to the monetary slowdown. Inflation clearly did not react as rapidly as I expected that it would." "Chastened by this experience," he had now reexamined the evidence and found that changes in the money supply hit industrial production quickly but affected prices much more slowly—with a lag averaging twenty to twenty-three months. Such a lag before money-supply changes affected prices, he said, was "much larger than I expected." But politicians were impatient with the delay and had imposed controls rather than wait for monetary policy to take effect.

"We have been driven into a widespread system of arbitrary and tyrannical control over our economic life," Friedman said, "not because 'economic laws are not working the way they used to,' not because the classical medicine cannot, if properly applied, halt inflation, but because the public at large has been led to expect standards of performance that as economists we do not know how to achieve." Perhaps, he added, as knowledge advanced, economics could come closer to prescribing policies that would achieve higher standards of performance. Or perhaps, he worried, the "random perturbations inherent in the economic system" would make it impossible to achieve higher standards of price stability, high employment, and steady growth; or perhaps, even if there were policies that would attain such standards, considerations of what he called "political economy"—that is, politics—

would make it impossible for such policies to be adopted. "But whatever the future may hold in these respects," said Friedman, "I believe that we economists in recent years have done vast harm—to society at large and to our profession in particular—by claiming more than we can deliver. We have thereby encouraged politicians to make extravagant promises, inculcate unrealistic expectations in the public at large, and promote discontent with reasonably satisfactory results because they fall short of the economists' promised land."

Whether the chastening experience of the failures of economic policy in the Nixon administration would last or not—and indeed, there would be evidences of Friedman's return to bold and precise prognostications and overadvertised policy results in the years to come—his economic philosophy is most firmly grounded in its long-term perspective. In the long run, the market does provide the best results; any hardships imposed by the market are only transitional and are best endured. Unfortunately, it is those least able to endure the hardships of the market economy—the poor, the racial minorities, the uneducated or undereducated—who are always expected to endure them. But in the long run, according to the Friedman philosophy, even they will be better off.

This positive view of the market is based on Friedman's sense of the beneficence of capitalism in history. Not only was capitalism responsible for bringing material prosperity to mankind, but it was also a necessary precondition for human freedom. In the concurrence of capitalism and freedom lay the hope of the world.

In recent years, however, this optimistic vision has become somewhat clouded. Of course, Friedman has long been aware that few people share his belief that the great achievement of capitalism was not the accumulation of property and wealth but "the opportunities it . . . offered to men and women to extend and develop and improve their capacities."[42] In *Capitalism and Freedom* he lamented that those who had most to gain from capitalism were often its most strenuous detractors,

> those minority groups which can most easily become the object of the distrust and enmity of the majority—the Negroes, the Jews, the foreign-born, to mention only the most obvious. Yet, paradoxically enough, the enemies of the free market— the Socialists and Communists—have been recruited in disproportionate measure from these groups. Instead of recognizing that the existence of the market has protected them from the attitudes of their fellow countrymen, they mistakenly attribute the residual discrimination to the market.[43]

But this wrongheaded dislike of the market, Friedman now feels, does not constitute the only threat to capitalism. The greatest threat, he suspects, comes from that very freedom which the capitalist system has created. It is the Western democracies which have moved, by the free choice of the people, in the direction of socialism and the welfare state. This has, in his view, brought about severe inflation and a resultant economic chaos that may well lead to collectivist and totalitarian forms of government. The supreme irony is that many capitalist strongholds—Taiwan, South Korea, Hong Kong, Spain, Argentina, Brazil (whose system of price indexation Friedman has professed to admire), and Chile (whose leaders profess themselves to be admirers of Friedman's economic doctrines)—are themselves unfree states where civil liberties are daily infringed and destroyed. Was the Nixon administration, with its heavy capitalist ideology, not headed in the same direction? To be sure, Friedman has never maintained that capitalism is in itself sufficient to produce political freedom, but if capitalism is no guarantor of freedom, is socialism necessarily its nemesis?

John Kenneth Galbraith

Socialism without Tears

"For the mob is varied and inconstant,
and therefore if a Reputation is not
carefully preserved, it dies quickly."
—*Spinoza*

I

THE TROUBLE with becoming a celebrity is that it makes one seem like other celebrities, many of whom are cardboard figures that fade and yellow in a season or two. Economist, ambassador, best-selling author, price controller, Harvard professor, political iconoclast who can be simultaneously inside and outside the establishment, television talk-show star, *Playboy* and *New York Times* contributor, skier, wit, and bon vivant, John Kenneth Galbraith has suffered the denigration routinely imposed upon celebrities by intellectuals and, sooner or later, by the press. In the journalism review (*More*), Gerald Nachman, Mary McGeachy, and Randall Poe in 1974 offered "trend journalists" a brief guide to current reputations; under the heading, "Too-Much-of-a-Good-Thing Dept.," they listed Levis, Gene Shalit, Coca-Cola, transcendental meditation, the Sunday *New York Times*, Johnny Cash, Judge Crater, Solzhenitsyn, John Kenneth Galbraith, Szechuan cooking, the Mafia, Eric Sevareid, tennis, Baskin-Robbins, Beverly Sills, Howard Cosell. . . .[1]

Galbraith needs and deserves rescue from his celebrity, even from his own charming version of himself. He is, first and foremost, an economist, and a serious one. For Galbraith, as for other leading professional American economists, the Great Depression of the 1930s was a seminal experience; it invested economics with a significance for the whole of social and political life that it has never lost. When the Depression descended upon North America, John Kenneth Galbraith was an undergraduate studying agricultural economics and animal husbandry at the Ontario Agricultural

College at Guelph. He came from a moderately well-off set-tlement of Scotch-Canadian farmers on the north shore of Lake Erie. The Galbraiths were one of the tallest clans in Southern Ontario, literally "pillars of the community."

Even in Canada, the Scotch were a thrifty people, and proud of it; the University of Guelph was inexpensive but not a very good school, according to Galbraith. "Leadership in the student body was solidly in the hands of those who com-bined an outgoing anti-intellectualism with a sound interest in livestock. This the faculty thought right."[2] Young Gal-braith was not happy there. In his senior year he applied for and won a research assistantship at the Giannini Foundation of Agricultural Economics at the University of California at Berkeley.

In 1931, at the age of twenty-three, Galbraith crossed the border to seek his fortune in the United States. Berkeley was a revelation. Professors like Henry Erdman and Howard Tolley actually encouraged their graduate students to think and question, and long discussions were the norm. As I have said elsewhere,[3] the economist's mode of thought develops out of what I would call his quasi-Talmudic training, which is long on discussion and debate, with continuous passage from the specific to the general and back again, savagely close in its textual criticism, skeptical about its own or anyone else's results, and complicated and wide ranging in its style of in-quiry. My own teachers, Professors Calvin Bryce Hoover and Joseph Spengler, impressed upon me that economics could never be a monologist's art, that the economist always needed to try out his reasoning on some other economist. This prob-ably explains why the economics profession is so strong and close a fraternity; the economic monologist is not just out of touch but always in danger of becoming a crackpot or, less seriously, a layman, if he cannot talk, talk, talk with his brethren—or, at least, read, read, read from them and to them. Preferably both.

But the fraternal way that economists learn and practice

their art has much to do with the failure of economists to do an effective job in educating the public. For economists talk mostly to each other. They tend to regard the public (even ex-economists or ex-academic economists or even ex-economics department economists rather than business school economists) as beyond the pale. Exclusivity exists in every profession, of course, and it is, if properly exercised, a good thing. As Everett C. Hughes put it:

> Every profession considers itself the proper body to set the terms in which some aspect of society, life or nature is to be thought of, and to define the general lines, or even the details of public policy concerning it. . . . These characteristics and collective claims of a profession are dependent upon a close solidarity, upon its members constituting in some measure a group apart with an ethos of its own. This in turn implies deep and lifelong commitment. A man who leaves a profession, once he is fully trained, licensed and initiated, is something of a renegade in the eyes of his fellows. . . . It takes a rite of passage to get him in; another to read him out. If he takes French leave, he seems to belittle the profession and his former colleagues. . . .[4]

John Kenneth Galbraith, a mischievous man, was to make a high art of living both inside and outside the economics profession.

I I

AT BERKELEY, as his rite of passage, Galbraith was marched through Alfred Marshall's classic *Principles of Economics* by Ewald Grether. He studied economic history with M. M. Knight and read Veblen, the institutionalist, sociologist, and homespun heretic from the orthodoxies of the American way of life.

Though it was years before the Keynesian revolution really

broke on American campuses, Leo Rogin was discussing Keynes in Berkeley "with a sense of urgency that made his seminars seem to graduate students the most important things then happening in the world."[5] Galbraith's fellow graduate students were as exciting as his professors. "The most distinguished," he later said, "were Communists," the rest "uniformly radical."[6] But Galbraith, just off the farm and an agriculturalist to boot, felt ideologically backward—fascinated by but not willing to join the radicals.

He earned his $720-per-annum research stipend by trying to discover consumer preferences for various sorts of honey. To do his work on honey, he needed to concern himself with the reasons for the depressed prices of the California produce market. In his third year at the University of California, he was sent to Davis, then a small college devoted almost exclusively to research and instruction in farming. There he taught most of the courses in economics and finished his doctoral dissertation.

But Galbraith's love affair with the University of California came to an abrupt end. Shortly before receiving his doctorate in the spring of 1934, he was offered an instructorship at Harvard. Hoping to use the offer of a Harvard post and the proffered $2,400 salary as bargaining material at California, he approached the dean of the College of Agriculture, only to make the horrified discovery that he was firmly expected to accept the post. And so he did.

In Cambridge, Galbraith was promptly installed, under the newly organized Harvard residential system, as a tutor in Winthrop House. There he became a friend of Joseph Kennedy, Jr., and later met his young brother Jack, the future president.

Galbraith had been hired to teach agricultural economics at Harvard. But this relatively circumscribed area of economics seemed woefully inadequate to explain, let alone ameliorate, the condition of farmers during the 1930s. The agricultural problem demanded a broader perspective. The classical theory of supply and demand maintains that a decline in the demand for a good will cause its price to fall; so, too, a general fall in demand, resulting from a declining business cycle, should cause prices generally to fall. But in the Depression, though farm prices fell sharply, industrial prices —the farmer's costs for fertilizer, machinery, and other goods—remained remarkably rigid. Here was the cause of the farmer's plight.

Galbraith had seen at first hand how declining farm prices had made a necessity of virtue among the parsimonious Scotch. He had studied the depressed state of agriculture in California. For him, as an agriculturalist, the need was clear: he must explain the persistence of high farming costs in the face of a general fall in aggregate demand and farm prices.

He was not alone in focusing on the problem of industrial price rigidities. The subject provoked intense study and debate in the early years of the Depression before Keynes's magnum opus, *The General Theory of Employment, Interest, and Money*, arrived in 1936 to dominate economic discussion. Galbraith's first contribution to the discussion was a paper entitled "Monopoly Power and Price Rigidities."[7] Already in this early essay of 1936, the prime concerns and distinctive hue of Galbraithian economics emerged. He took his cue from the exciting works on market theory and the structure of industry that had just appeared: E. H. Chamberlin's *Theory of Monopolistic Competition*, published in 1932, and Joan Robinson's *The Economics of Imperfect Competition*, published in 1933. Although Chamberlin, a pupil at Harvard of the great economist Allyn A. Young, was .to spend decades marginally differentiating his book from Robinson's, both works forcefully demonstrated the need to expand the traditional bipolar classification of markets, including pure competition (a large number of firms selling undifferentiated products against one another) and monopoly (one seller facing no direct competition), into new intermediate models involving a few sellers well aware of each other's "monopolistic" or "imperfect" competition for markets in which goods were partially differentiated by brand name, style, design, or other characteristics. The sellers in such markets might in fact spuriously differentiate their products by advertising and salesmanship. Each seller was held to control a large enough share to set his own prices, influence the overall level of prices in his field, and affect total supply.

In 1932, Adolf A. Berle and Gardiner C. Means had published another influential book that put meat on the Chamberlinian and Robinsonian bones; their book, *The Modern Corporation and Private Property*, a study of the proportion of national wealth, industrial wealth, and corporate assets owned by the 200 largest nonfinancial corporations in the United States, demonstrated that the industrial sector of the American economy was narrowly concentrated. A couple of hundred corporate entities owned over half the nation's assets employed in manufacturing, transportation, utilities, and mining. Further, there had developed a cleavage between the nominal owners of business corporations, the stockholders, and the actual controllers of industry, the "hired" managers; it was the managers who had become the real oligarchs of American industry.

Galbraith found the new theory of oligopoly and monopolistic competition compelling. He had himself studied the role of prices and costs in the industrial versus the agricultural sector.[8] The Berle-Means study reinforced the importance of the newly recognized market forms: with gargantuan corporations dominating the industrial landscape, the American economy was only very imperfectly competitive. While there were few strict monopolies, industrial market power was the norm, not the untoward exception.

This, Galbraith argued, was the essential explanation for the rigidity of industrial prices, and hence for the sickness of agriculture. In truly competitive markets, which the farmers faced, a producer has no choice but to sell what he has at the impersonally determined market price. He does not himself control enough of the market to affect the price by withholding his supply or dumping it all at once on the market. To offer his goods—indistinguishable from anyone else's—at other than the market price would mean either throwing away money (if he sold below the market price) or doing without customers (if he tried to sell above it). But the monopolistic competitor can maintain or even raise his prices, confident that at least some of his customers will stay with him. Or he can cut his prices to gain a larger market and expand his profits by driving out weaker competitors.

Producers with strong market power can thus set their own prices and dominate overall price levels in their respective industries. Galbraith sought to show that producers with monopoly power might hold prices high even when there was a shortage of aggregate demand, although this might mean sacrificing short-term profits. First, he argued, oligopolists have most to fear from a price war. It is in their interest to adjust to decreased demand by reducing production rather than lowering prices, despite the cost of holding on to a large inventory. In these circumstances, an oligopoly (an informal league of important sellers) would tend to function like a monopoly, eschewing short-run profits in the interest of market stability. Second, the monopolistic competitor tends to keep prices constant to encourage brand loyalty:

> If the differentiation is largely psychic, as the result of advertising, the consumer is educated to ignore brands as price attractions lead him to shift from one to another which he finds equally good. The differentiation is broken down and a part of the advertising expenditure is wasted.[9]

Sales may fall, but the producer can again solve the problem by cutting production. Finally, the system feeds on itself.

Since industrial prices in one area of production are costs in another, the whole structure of industrial prices holds itself up. Thus, techniques of price and production control cushion the entire industrial sector from the harsh effects of reduced demand.

The farmer had no such cushion during the Depression. Galbraith held that the economy had shown a tendency "to break open along an ill defined line of cleavage between competitive production and production characterized by one or another form of monopoly power."[10] What should be done about this? Galbraith offered some hints on how to proceed. Attitudes towards monopoly power should be revised; in industry it was not a special case but the normal state of affairs. Moreover, antitrust legislation could not be seen as a cure-all, since this merely substituted oligopoly for monopoly. He urged, rather vaguely, that prices and price levels be harmonious throughout the economy—that is, on both sides of the cleavage. In addition, he argued, a new look should be taken at copyrights, patents, and brand advertising, to reduce the market power of individual firms. Likewise, a "positive program for extending consumer standards" should be instituted, going well beyond the pure food and drug laws and other existing government regulations. Galbraith was concerned that his fellow economists bring their thinking up to date with the realities of a new, noncompetitive economy:

> Political action has moved ahead rapidly on these matters. The economist is scarcely justified either in condemning it unless he has answers to the questions just raised.[11]

This early paper is interesting because it contains the central themes present in all of Galbraith's later work: the division of the economy into purely and imperfectly competitive sectors, a focus on situations where pure competition and a strict profit motive are lacking, a concern with consumerism and the detrimental effects of advertising, and the desire to evangelize the economics profession. Moreover, Galbraith here employed the same method of argumentation which characterizes his later works, though more explicitly expressed and more modestly assessed:

> The foregoing pages are essentially an exercise in qualitative analysis. The understanding of price rigidities is scarce begun until progress has been made in weighting or quantifying the above mentioned influences as they appear in the actual process of industrial price making.[12]

The task of quantification never much appealed to Galbraith. His approach has remained nonempirical, synthetic, impressionistic.

III

IN THE FALL of 1937, Galbraith, newly married to Catherine Atwater, left Harvard to spend a year as a social science research fellow at Cambridge University. This was a productive time for him. He collaborated with the manufacturer H. S. Dennison on *Modern Competition and Business Policy*, an early effort to elaborate a Galbraithian portrait of the industrial system.[13] He described the imperfection of the marketplace and asserted that its inability to adjust appropriately to decreased demand meant that the American economy had largely ceased to be self-regulating. Antitrust legislation could have little effect on such rigidity. A program of industrial regulation was, however, adumbrated; its purpose would be to increase production hampered by monopoly power in the marketplace.

By the time the book appeared in 1938, however, the Keynesian revolution was rocking the economics profession. It rendered Galbraith's observations, for the nonce, trivial.

Market imperfections might or might not create price rigidities and make the farmer miserable; they might or might not hold back production and hamper employment. The economy was not, indeed, self-regulating, for equilibrium could as well be reached at a low as at a high level of unemployment. The solution, however, was for the government to put money into people's pockets by cutting taxes and spending more. Then, with increased aggregate demand for goods and services, farm prices would rise and industry would expand production, with the consequent spur to employment. The old downward rigidities would become a faded memory of the unenlightened past; prices throughout the economy would again be "harmonious." The alternative course of structural reform was likely to be long, tortuous, and politically uncertain.

The Keynesian program seemed almost too good to be true. Galbraith subscribed and joined in the attempt to spread the good word. He ghosted a Keynesian manifesto, directed at the business community, which appeared under the names of Dennison, Morris E. Leeds, Ralph E. Flanders, and Lincoln Filene, all liberal New England businessmen.[14] Returning to Harvard in the fall of 1938, Galbraith found that Alvin H. Hansen, who was to become known as "the American Keynes," had been installed as professor of political economy. A conventional and orthodox economist during his earlier years at the University of Minnesota, Hansen became the doyen of the brilliant Keynesian circle at Harvard which included, among others, Paul Samuelson, Alan Sweezy, and Walter Salant. Professor Hansen, a key New Deal adviser, married economic theory to politics:

In the late thirties Hansen's seminar in the new Graduate School of Public Administration was regularly visited by the Washington policy-makers. Often the students overflowed into the hall. One felt that it was the most important thing currently happening in the country and this could have been the case.[15]

Never before and never since have economists felt so vivacious. "Bliss was it in that Keynesian dawn to be alive, but to be young was very Heaven!" At a time of the most severe and widespread economic hardship, they possessed the cure, the true philosopher's stone. Moreover, in restoring prosperity and saving capitalism from itself, the young Keynesians could have all the fun of being revolutionaries. The Marxists would be upstaged, exposed as the dupes of apocalyptic dreams and passé economic theory.

The academy, however, offered little scope for revolution, and many headed for Washington to put conviction to the test. Galbraith felt the pull. Working with G. G. Johnson, Jr., a Treasury Department economist and recent Harvard Ph.D., Galbraith undertook a study of New Deal public works programs for the National Resources Planning Board. This study, "The Economic Effects of the Federal Public Works Expenditures," appeared at the end of 1940; it recommended the establishment of a long-range national public works policy for the dual purpose of furnishing employment during a depression and providing a program of public construction at ✓ all times. Meanwhile, Galbraith's Harvard instructorship had run out, and he had accepted the post of assistant professor at Princeton. In 1940 he became an economic adviser to the National Defense Advisory Commission, and in 1941 he left Princeton to head the Price Section of the Office of Price Administration and Civilian Supply (OPA).

From the standpoint of bureaucratic aggrandizement, Galbraith was in the right place at the right time. After Pearl Harbor, price controls and rationing became the order of the day. Under the benevolent eye of the prominent New Deal economist and chief of OPA, Leon Henderson, Galbraith, then thirty-three years old, saw his staff expand from a dozen to 16,000 in a year and a half. These minions included several of his old professors at Berkeley, as well as a young lawyer from southern California named Richard Milhaus Nixon.

The experience of World War II validated Keynes. Deficit spending at an unprecedented level was employed by the government to purchase war matériel and finance the armed services. To meet the orders, American industry hired the additional workers needed for production at full capacity. With increased demand for industrial goods augmented by the manpower requirements of the Army, unemployment disappeared. Consumers had more money to spend than they had had in a decade. Inflation replaced depression as the problem. The increased demand for consumer goods bid up prices, and as firms expanded production to meet the demand, needed resources were in danger of being drawn away from war production. This had to be avoided at all costs. Hence, the dual role of the price administrator was to check inflation and help in the larger strategy of resource mobilization.

The price administrator has a thankless task; his every decision must make someone unhappy. Galbraith, however, was fortunate in one respect. Although he had his share of political and bureaucratic conflict, his economist's conscience was untroubled. Most economists—Keynesians included—regarded price controls as at best a necessary evil, reserved for times of all-out war. A substantial body of opinion held that controls both could and should not work; dire consequences were predicted to follow from such untoward tampering with the smooth operation of the marketplace. Galbraith did not agree. His observation had been that, for much of the economy, monopoly power had already impaired the market's price-setting and allocative mechanism. Prices were already being administered by many producers. There would be no change in how many of the prices were set, but only in who did the setting. And in some areas a government agency with the public interest in mind could do a better, more efficient job.

As head of price administration for OPA, Galbraith felt he had found confirmation for his picture of the American economy. Price control proved to be much easier to enforce in imperfect than in more purely competitive markets. OPA had little trouble policing the markets for primary metals and retail milk; in both cases, a few large sellers were in command. However, in the markets which approached pure competition, such as those for scrap metal and fresh vegetables, prices were almost impossible to control. The explanation for this phenomenon was not, Galbraith felt, merely that it was easier to watch a smaller number of firms. Rather, there was a fundamental difference in pricing mechanisms. Where the free market operated to set prices, controls were, perhaps of necessity, ineffective. But as Galbraith later wrote, "It is relatively easy to fix prices that are already fixed."[16]

During his career as a wartime price controller, Galbraith also obtained—at the cost of considerable personal anguish —confirmation of his belief in the power of big business and in its natural alliance (which could become an unholy alliance) with government. The price controllers had taken on a heavy assignment, which brought them into conflict with business and political forces, even during the great patriotic war. To combat the wartime inflation resulting from excess aggregate demand, the government might have put heavier reliance on monetary and fiscal restraints but was inhibited from doing so. The objective of bringing the economy quickly up to full production prevented extensive use of monetary restraints. And fiscal policy was inhibited because it was deemed unfair to tax too heavily a population that was being asked to work overtime and contribute in many other ways to the war effort. Confiscatory taxes might have undermined public morale; hence, far greater resort was made to voluntary savings and war-bond programs. As a result, price administration became the principal weapon for combating inflation, a weapon impaired by the exemption from control of raw agricultural goods, largely because of pressures from the farm lobby.

Despite the odds against it, OPA took its job of holding the line seriously. In April 1943, President Roosevelt made a speech stressing the need to restrain inflation. In May, OPA announced plans to cut consumer prices for meats, butter, and coffee about 10 percent; food processors would be compensated by government subsidies. In addition—and this was particularly close to Galbraith's heart—a program of brand standardization and quality stamping would be instituted on consumer dry goods, with price ceilings established on the various quality levels. In an assessment of the program, Turner Catledge of the *New York Times* wrote:

> As for its part in the anti-inflation fight, the OPA is in the position of one unit in a coordinated military attack. This forward force has run far beyond the others that were supposed to advance with it. As a result it is standing out there with its flanks exposed, taking a terrible bulk of fire, while the other authorities are even arguing over whether they will advance or not.[17]

The heaviest fire came from the food industry and the dry goods industry. Galbraith was singled out for attack. Giving testimony before the House Interstate Commerce Committee, Lew Hahn, general manager of the National Retail Dry Goods Association, brandished a copy of *Modern Competition and Business Policy,* in which he had discerned a plot to suppress brand names in order to destroy the "price jurisdiction" of the manufacturer. This, Hahn claimed, was clearly the philosophy behind OPA orders. He accused Galbraith, not entirely unfairly, of trying to "change the business structure of the nation under the guise of war necessity."[18]

Meanwhile, Galbraith had also become vulnerable to attack within OPA itself. In January 1943 a former Michigan Senator, Prentiss Brown, had taken over from Leon Henderson as chief of OPA. As his assistant, Brown installed Lou Maxon, who as the head of a large Detroit advertising agency personified for Galbraith the twin evils of powerful oligopoly and salesmanship. Maxon, on his side, made no secret of his hostility to Galbraith and to "the professors and theorists whom Dr. Galbraith represents."[19]

A power struggle was on at OPA, and the outside criticism of Galbraith was perhaps not entirely coincidental. The dustup gave new life to congressional desires to gain some power over price administration. Congress had to be appeased; Maxon was appointed deputy administrator of OPA, and Galbraith resigned. Persuaded that Maxon was more practical and less incorruptible than Galbraith, businessmen and politicians scaled down their opposition.

I V

DURING the succeeding years, Galbraith alternated between government service, journalism, and academia. After leaving OPA, he attempted to enlist in the Army, but was rejected because of his height. (He is 6 feet 8 inches tall.) He was named to the staff of the Lend-Lease Commission in August 1943, but he left shortly thereafter to join the editorial board of *Fortune* magazine. While at *Fortune*, Galbraith wrote some fifty articles, editorials, and book reviews. He has often claimed it was there that he learned how to write. This is an exaggeration. Galbraith's pre-*Fortune* prose is clear and straightforward; what he may have acquired from Mr. Luce was a taste for phrasemaking and a certain archness of style. Shortly after Germany surrendered in May 1945, Galbraith was named a director of the U. S. Strategic Bombing Survey. The purpose of the survey was to determine the effect of Allied bombing on German war production. Aerial reconnaissance and some wishful thinking of the part of the U.S. Army Air Forces suggested that the bombing had been devastating to German industry, yet the material well-being of the German Army and Air Force suggested the contrary. What was needed was the Third Reich's production figures. The search for a comprehensive picture of the German war economy led the USSBS investigators to Albert Speer, the canny architect who had risen in the Nazi hierarchy to become manager of production. Galbraith assisted at a series of interrogations of Speer which were conducted by George Ball,[20] another director of the USSBS, later to become undersecretary of state and famous as an opponent of the Vietnam war during the Johnson administration.

In the course of its work, the USSBS compiled the first competent figures available on Germany's gross national product. In the end, the conclusion was unavoidable: the Allied bombing, conducted at great cost, had not significantly impaired German war production. In some cases, the bombing may even have spurred production, for where cities were struck, out-of-work civilians—waiters, shopkeepers, and the like—were impelled to seek employment in often underutilized industrial plants. This lesson in the dispensability of much ordinary economic activity did not escape Galbraith.

In 1946, Galbraith was appointed director of the State Department's Office of Economic Security Policy, which was responsible for overseeing economic policies in the occupied countries. The responsibility was nominal, however, given the twin hegemonies exercised by Generals Lucius Clay in Germany and Douglas MacArthur in Japan. For his pains, however, Galbraith received the Medal of Freedom. The next year saw Galbraith back at *Fortune*, and 1948 marked his return to academe. After a year as lecturer, he was in 1949 made professor of economics at Harvard, where he remained, except for leaves, happily ever after.

Harvard gave him both the forum and the freedom he wanted. During the McCarthy period in the early 1950s, Galbraith escaped the wrath of congressional investigators while many of his old radical Berkeley friends did not. For this he could thank his earlier political reticence and Harvard President Pusey's courageous support of his faculty.

Galbraith's academic career proceeded apace. In 1952 he published two books—a slender volume entitled *A Theory of Price Control,* and a somewhat stouter one, *American Capitalism. A Theory of Price Control,* an expansion of two previously published articles,[21] is part of a modest tradition of apologetics written by economists who have seen official service as price controllers. The genre was initiated by Professor F. W. Taussig, a member of the Price-Fixing Committee in World War I, in a famous paper "Price-Fixing as Seen by a Price Fixer."[22] A recent contribution is "Making Wage Controls Work," by Professor Arnold Weber, former head of President Nixon's Pay Board.[23] Galbraith's book appeared in the context of the rapid inflation at the outset of war in Korea. His purpose was to convince fellow economists to regard price controls as a viable and effective tool of economic policy. Even in 1952 this was not an easy task.

Galbraith began by stating his conviction that circumstance, not high-level theory, determines the course of policy. "It is part of man's pride that he makes economic policy; in fact, in economic affairs, he normally adjusts his actions, within a comparatively narrow range of choice, to circumstances."[24] Such had been the case with the comprehensive system of price fixing in World War II, established under the General Maximum Price Regulation ("General Max") of 1942. "Events had forced the step that economists, in the main current of economic theory, had so long viewed as unwise or impossible, or both."[25] The policy dictated by the circumstance of World War II had, Galbraith asserted, worked rather well. It was important to examine the reasons for this perhaps surprising success. Why? Because inflation could no longer be regarded as a transient or wartime phenomenon. "Inflation, more than depression, I regard as the clear and present economic danger of our times and one that is potentially more destructive of the values and amenities of democratic life."[26]

He invoked his old characterization of the nature of modern industrial markets as the key to the success story of wartime controls. Admittedly, price controls would have been unworkable—as economic theory held—in a purely competitive market. But pure competition did not reign in the American marketplace. Monopoly power impinged as a matter of course on the innocent interaction of supply and demand. Monopolists, oligopolists, and monopolistic competitors set their own prices, and these tended to be inflexible, even institutionalized in varying degrees. Retailers' markups were largely standard, unaffected by changes in supply and demand. Thus, both existing business price administration and customary business practices made it easier for the government to exercise control. Other phenomena—the time lag between imposition of a control and shortage of a given good, the ability of industry to increase production at constant or decreasing unit costs—would also help government to check inflation with only minimal resort to rationing.

Yet inflationary pressure was bound to persist because, at bottom, management of the wartime economy entailed perpetuating an excess of aggregate consumer demand for the sake of war production. No amount of market imperfection could alter this reality, which Galbraith entitled "The Disequilibrium System." The success of the Disequilibrium System was measured by increased output and depended ultimately on the dedication of the American people. War needs required that a greater amount of work be done than was necessary to purchase the available supply of consumer goods. The willingness of Americans to work hard and save added income, and not to work merely to satisfy immediate personal needs, was responsible for the miracle of U.S. war production.

The vast excess of aggregate demand built up in the course of World War II was unleashed when, in 1946, a Republican Congress swept away all price controls. A burst of inflation followed. Galbraith was sharply critical of the decision, holding that controls should have been relaxed gradually.

> As so often happens in American life, the essentially radical course was advocated by conservatives. In this instance the course, designed to weaken capitalism by detracting from the integrity of the promise to the nation's debtors and to force patterns of economic behavior—violent price movements and concomitant industrial strife—that are least easy to defend, was pressed by those who count themselves capitalism's most ardent protectors.[27]

Galbraith's prime concern was not, however, to score points against conservatives for the past but to teach a lesson for the present and future. In the late 1940s and 1950s, economists had begun to talk about what they called the "wage-price spiral," inflation accelerated by the interaction of wages and prices in the industrial sector. Labor unions would bargain for wage increases somewhat exceeding the existing rate of inflation, in order to catch up with past price increases and anticipate future increases. Businesses would grant the increases, confident that, with demand for their products high, the added labor costs could be passed along to consumers in the form of higher prices. These higher prices would in turn generate bigger wage demands, and the cost of living would continue to rise indefinitely.

Galbraith was much impressed by the wage-price spiral. It reinforced his belief in the imperfection of industrial markets;

the economic power of business and labor was clearly adding to the whole society's burden of inflation. To be sure, the wage-price spiral depended on the existence of high aggregate demand, which kept the sales of most goods from being hurt by a rise in their prices. With plenty of money in their pockets and bank accounts and ready access to credit, consumers were willing and able to absorb higher prices. With aggregate demand low, however, industry would be unable to raise prices with impunity, since sales would then drop too sharply. Hence, business would have to adopt a tougher line with the unions. The wage-price spiral would wind down. The obvious solution was to lessen aggregate demand to stop the inflationary wage-price spiral.

Nevertheless, Galbraith predicted a long-term continuation of high demand. This he attributed, back in 1952, to the country's defense needs:

> The sort of limited mobilization on which the United States has recently been engaged has no foreseeable terminal point. The accepted view of the political and military strategy underlying American rearmament is that an effective and modern military force in being is necessary for peace.[28]

Inflation would, of course, disappear if there was sufficient slack in the economy—less than full employment and production. But, argued Galbraith, recourse to such slack was prohibited by the exigencies of national defense, for supplying the military would entail full use of resources. Was inflation, then, inevitable? Perhaps, answered Galbraith, but it could be kept within bounds.

Price administration by government had an important role to play in any program of inflation control. While fiscal and monetary policies should be used to establish as close a balance of supply and demand as possible, price control would still be needed to arrest the wage-price spiral.

> In an ideal model of inflation control, when the requisite restraint is being kept on demand, price controls would be con-

fined to imperfect markets where prices are administratively determined. Wage controls would be confined to those wages that are set by collective bargaining with effective unions.[29]

Wage and price controls were thus only to be used for correcting the inflationary distortion caused by exercise of economic power. Fortunately, though hardly by coincidence, controls would have to be applied precisely where they were easiest to administer. Market power—which made the wage-price spiral possible—itself facilitated public administration of wages and prices. All that was needed was public understanding that, in the post-Keynesian era, inflation would remain the most serious threat to economic stability, persisting even during recession. Immense military expenditure would be no transient phenomenon; the threat of the wage-price spiral was here to stay.

V

A THEORY OF PRICE CONTROL went largely unnoticed. Galbraith found this lack of recognition from his peers distressing; he has since called it the reason for his decision to refrain from writing for economists alone:

> I made up my mind that I would never again place myself at the mercy of the technical economists who had the enormous power to ignore what I had written. I set out to involve a larger community. I would involve economists by having the larger public say to them, "Where do you stand on Galbraith's idea of price control?" They would *have* to confront what I said.[30]

Of course, Galbraith had long been concerned with writing for a larger audience. His collaboration with Dennison, his Keynesian tract ghosted for the liberal business leaders, and his work at *Fortune* had all been outside the confines of academic economics. Moreover, his first best-seller, *American Capitalism*, had actually come out shortly before *A Theory of Price Control*.

Nonetheless, it is important to consider why Galbraith has since couched his contributions to economics in the form of best-selling nonfiction. The desire for fame and fortune has no doubt played a part, but this must be taken as a necessary but not sufficient explanation. In fact, writing for the public lies at the heart of Galbraith's approach to economics.

In Plato's *Gorgias*, Socrates distinguishes between belief and knowledge. Belief can be true or false, but knowledge must by definition be true, since there is no such thing as false knowledge. Persuasion is used to establish both; however, as Gorgias the rhetorician admits, that form of persuasion which is rhetoric concerns itself *only* with securing belief.[31] Rhetoric is today a dirty word, and Plato's case against it need hardly be made. But the issue of knowledge and belief remains of enduring interest. Every scholarly field must postulate—whatever the philosophical hazards of such a position—the possibility of knowledge, of established truth. Thus, the scholar must always aim at adding to the stock of knowledge; he cannot seek merely to convince. This is the necessary fiction to which scholarship adheres.

The outside world is but little affected by the nature of belief and by the unwitting or unscrupulous means of securing it in most learned disciplines. Economics, for better or worse, is not one of these. People may be prosperous and happy, or suffer poverty and anxiety, according to the state of economic belief. This point of intersection between circumstance and belief is the true focus of Galbraithian economics. At a time when economics has sought to become more and more scientific, placing brick upon brick of securely quantified knowledge, Galbraith has approached the edifice as a system of belief, increasingly isolated from knowledge of the real world. His goal has been to show how such belief has diverted attention from economic realities and misdirected economic policy.

The opening chapter of *American Capitalism* is entitled "The Insecurity of Illusion." In it Galbraith noted a deep concern on the part of both liberals and conservatives about the nature of the American system. Liberals were troubled by the size and power of big business; conservatives, by the size and power of the national government. Business power was expected to exploit workers; government power, to destroy private enterprise. In fact, however, the system was functioning rather well. Workers were well off and profits were good. The economy was turning out goods and services at an unprecedented level. How, then, to explain the expressions of concern? The trouble lay not with the world but with the ideas by which it was interpreted. Those ideas were "the source of the insecurity—the insecurity of illusion."[32] 9

The illusion consisted of belief in the economic doctrine of pure competition:

> Men cannot live without an economic theology—without some rationalization of the abstract and seemingly inchoate arrangements which provide him with his livelihood. For this purpose the competitive or classical model had many advantages. It was comprehensive and internally consistent. By asserting that it was a description of reality the conservative could use it as a justification of the existing order. For the reformer it could be a goal, a beacon to mark the path of needed change. Both could be united in the central faith at least so long as nothing happened to strain unduly their capacity for belief.[33] 17

But belief had been strained. Large businesses set their own prices and bargained collectively with large labor unions. The federal government took an active role in managing the economy. The competitive model jarred distressingly with reality.

Yet why were people so loath to abandon the notion of pure competition? Galbraith advanced a dual explanation. Economists liked it because it solved the problem of efficiency. Embodying a mechanism to ensure the most efficient allocation of resources, it thus made possible achievement of the economist's highest goal: the alleviation of poverty. For political philosopher and businessman alike, the competitive model solved the problem of power. In a world of small firms, no one possesses enough economic power to force his will on anyone else; all are equally subject to the impersonal forces of supply and demand. Democracy is the handmaiden of the purely competitive system. Likewise, there is no personal responsibility for economic hardship. Small wonder that much noble purpose and much self-interest should attach to pure competition.

Galbraith's purpose in *American Capitalism* was twofold. First, he sought to show that the competitive model was neither a description of reality nor a viable reforming ideal. Second, he wanted to construct a new theory of the American economic system, as internally consistent as the old, but which adequately reflected the real world. The new theory would allay the fears of liberals and conservatives alike, and enable Americans to accept their system for what it was.

Galbraith already had a description of the American economy in hand. For fifteen years he had been preoccupied with the existence of monopoly power in the industrial marketplace. In *American Capitalism*, Galbraith displayed this power to the public. The United States, he asserted, was characterized by a vast degree of economic concentration; to imagine that such concentration could be extirpated by antitrust prosecution was absurd.

To suppose that there are grounds for antitrust prosecution wherever three, four or a half dozen firms dominate a market is to suppose that the very fabric of American capitalism is

John Kenneth Galbraith

illegal. . . . The liberal, who still searches for old-fashioned monopoly in the modern economy, has been made to feel that his is a search for poison ivy in a field of poison oak.[34] 55

Yet all was not so bleak as theory had suggested. The economy was healthy and expanding. Galbraith, in fact, was sanguine about some aspects of monopoly power. Borrowing the hypothesis of Professor Joseph Schumpeter, he argued that a monopoly, having funds available for research and development, was particularly well suited to innovation, the lifeblood of capitalism. New technologies developed by those possessing monopoly power more than offset the detrimental effects of market imperfection. Price collusion by oligopolists, much deplored by economists, was the necessary alternative to self-destruction, Galbraith held. The salesmanship and advertising which resulted from the rejection of price competition was a benign, not a malignant, tumor.

> Our proliferation of selling activity is the counterpart of comparative opulence. Much of it is inevitable with high levels of well being. It may be waste but it is waste that exists because the community is too well off to care.[35] 96

A broader theory was nonetheless needed to replace pure competition and explain the success of the American system. This Galbraith developed under the name of "countervailing power." The theory explained how the U. S. economy was kept on an even keel despite the all-pervasive existence of concentrated power:

> Private economic power is held in check by the countervailing power of those who are subject to it. The first begets the second. The long trend toward concentration of industrial enterprise in the hands of a relatively few firms has brought into existence not only strong sellers, as economists have supposed, but also strong buyers, as they have failed to see. The two develop together, not in precise step but in such manner that

[121]

there can be no doubt that the one is in response to the other.[36] \\\

Thus, strong labor unions emerged in areas of high industrial concentration, such as the steel and automobile industries, but did not emerge where production was more purely competitive, as in farming. Chain stores and mail order companies grew up in response to monopoly power in the production of consumer goods. In this way, exploitation of the weak by the economically powerful was counteracted. A balance of power restored efficiency to markets made imperfect by monopoly control.

Of course, countervailing power was not present everywhere. In the housing industry, small contractors were squeezed between strong labor unions and oligopolistic suppliers. Farmers producing for competitive markets had to buy equipment from concentrated producers. Such phenomena, however, far from weakening Galbraith's case, actually seemed to him to make it more comprehensive. It provided the government with a meaningful and legitimate role in the marketplace: helping the development of countervailing power.[37] Agricultural subsidies, for example, were no longer to be regarded as an aberrant exercise in economic and political expediency; rather, they represented a legitimate governmental extension of countervailing power to farmers.

A meaningful place could finally be vouchsafed to antitrust legislation. The relevant question was not "What constitutes monopoly power?" but "What impairs countervailing power?" Mergers and strict monopolies could still be resisted on the grounds that this provided a greater opening for the exercise of countervailing power. But effort would not be wasted in seeking to prevent actions which simply constituted a countervailing response to preexisting agglomerations of economic power.

Galbraith found but a single flaw in the theory and practice of countervailing power. This lay in inflation control. In times of excess aggregate demand, the various countervailing forces no longer checked but conspired with each other. The same mechanism which otherwise restrained the exercise of market power acted, under inflationary pressure, to produce the wage-price spiral. Labor and management could both increase their take, passing the cost on to the public.

The problem of inflation was the only cloud in the blue sky of *American Capitalism*. Galbraith could not handle it under the rubric of countervailing power. He insisted, however, that there was no completely satisfactory solution. The structure of the modern economy rendered the Keynesian program— so effective in dealing with the problems of depression— inadequate for handling conditions of boom. Restricting demand was a much trickier enterprise than expanding it. Galbraith, not surprisingly, suggested wage and price controls as a means of controlling inflation, but he did not consider these to be satisfactory in the long run. Capitalism, as he saw and approved of it, was "an arrangement for getting a considerable decentralization in economic decision."[38] Inflation, not depression, constituted the greatest threat to this arrangement. On a sober note, *American Capitalism* concluded:

> In any case there is no doubt that inflationary tensions are capable of producing a major revision in the character and constitution of American capitalism. Policy against depression, about which conservatives have been so deeply disturbed for so long, has little effect on essentials. Policy against inflation has a profound effect. Boom and inflation, in our time, are the proper focus of conservative fears.[39]   P. 201.

In *American Capitalism,* Galbraith sought to make his peace with the American system. The peace was not to be a lasting one; never again would he be so sanguine about the operation of the U.S. economy. Yet the book was no mere

[123]

apologia. In the midst of the Korean war and the McCarthy-
ite purges, Galbraith contributed an important new perspective to the sterile and destructive debate over capitalism,
socialism, and the American way. The public was well impressed. The notion of countervailing power legitimized activities of big business, big labor, and big government—those
vast and disturbing mammoths of American economic life.
Countervailing power, Galbraith implied, accorded easily
with that most comfortable of American dogmas, the federal
government's system of checks and balances.

Economists were less pleased with *American Capitalism*
than its lay readers. Some took Galbraith to task for writing
for the public at large; like Plato's rhetorician, he was accused of trying to make himself "appear better to those who
do not know than to those who know."[40] Too many and too
diverse phenomena had been treated as subject to countervailing power. Galbraith was charged with having skipped
too facilely from putative market imperfections to the total
performance of the economy. His documentary evidence was
woefully inadequate. By and large, the economics profession
attached to Galbraith's book that most damning of scholarly
evaluations: interesting.

But for the economists, the major problem with *American
Capitalism* was that it opened no doors to further work. This
troubled Galbraith as well. He had criticized his fellow economists for focusing excessively, under the influence of
Keynes, on maximizing the total output and employment of
the economy, while ignoring the inflationary dangers inherent
in a radically imperfect market structure. Yet here was the
difficulty, a seeming contradiction in Galbraith's thought. He
had argued in *American Capitalism* that conventional worries about the exercise of monopoly power were inappropri-
ate. Inefficiency attributable to administered prices or heavy
advertising and selling costs could be easily sustained by a
society untroubled by scarcity, especially since the large cor-

porations made economies of scale possible and were prone to innovate in both products and production techniques. Yet the theory of countervailing power, while holding that large accretions of economic power were essentially benign, offered no solution to the problem of inflation, which posed a genuine threat to the stability and even the survival of the system.

Where to go from here? Galbraith struggled for an answer. In 1954 he published *The Great Crash*, a chronological study of the stock market collapse of 1929.[41] This brilliant and stylish volume pointed up its author's concern about the dangers of speculative boom and bust. In 1955, Galbraith brought out *Economics and the Art of Controversy*, a revision of a series of lectures given the year before at the College of Puget Sound in Tacoma, Washington.[42] Here he argued that "the present topics of economic controversy have seen their best days."[43] Labor unions and collective bargaining were an accepted fact of life; so was government intervention in agricultural markets. The need for managing the economy along Keynesian lines—especially in warding off depression—was conceded. Both political parties agreed on the basic constitution of American capitalism. Not that all economic controversy had disappeared. Far from it. But economic debate had become largely pro forma, designed to procure tactical advantages within accepted institutional bounds. Thus, for example, "creeping socialism" could be invoked in an attempt to hold down Social Security benefits. Economics retained its capacity to provoke argument, but the government's actual policymaking options were increasingly limited:

No doubt there will be as much controversy over economic questions in the future as in the past, and economists should find this encouraging, for much of their income, as well as most of their prestige, derives from the persistent tendency of people to get exercised over this subject.[44]

[125]

For his own part, however, Galbraith was troubled by what he perceived as the lack of important economic issues. He was also troubled by the persistence of poverty in the richest country in the world. Perhaps the U.S. economy was not so well disposed as he had imagined. A book to be titled "Why People Are Poor" was projected, but the writing proved difficult. Galbraith sought a change of venue and from the summer of 1955 began frequenting the Swiss mountain resort of Gstaad. In 1958 he emerged with *The Affluent Society*.

V I

THE POINT OF DEPARTURE of the new book was once again the issue of belief and circumstance. Galbraith granted that common belief—which he termed the "conventional wisdom"—was a very powerful force.

> For a vested interest in understanding is more preciously guarded than any other treasure. It is why men react, not infrequently with something akin to religious passion, to the defense of what they have so laboriously learned.[45]

What finally makes the conventional wisdom obsolete is therefore not some new and revolutionary idea but a radical change in the reality which is the object of that belief. The person who calls attention to the new reality is wrongly credited with overthrowing the conventional wisdom. In fact, said Galbraith with his characteristic blend of ironic modesty and graceful arrogance, such an author "will have only crystallized in words what events have made clear, although this function is not a minor one."[46]

The purveyor of the new reality had of course to be specific about what elements constituted the old belief and where the old laws had been abridged. Galbraith provided a survey of economic thought from Smith to Marshall, in which he isolated three preoccupations: production, inequality, and insecurity. The evaluation of these issues was the enduring legacy of economics to the conventional wisdom, the heart of obsolete belief.

Until recently, Galbraith argued, societies—and hence economists—were most concerned about the problem of scarcity. Production did not outstrip by much the requirements for human survival. Economic thinkers rightly placed the highest premium on production at maximum efficiency. This task, it came to be seen, could be managed by the freely competitive market. Goods and services would be precisely allocated according to consumer demands. The efficient producer would be rewarded by greater sales; his competitors would be impelled to follow suit or fall by the wayside. The greater good of society was accordingly served. Unfortunately, said the conventionally wise, shedding a crocodile tear or two, the pursuit of efficient production entailed inequalities of wealth and economic insecurity. Riches were the reward of industry and efficiency; poverty was the evidence of (if not the punishment for) incompetence or sloth. The day-to-day uncertainties of businessmen and workers, the longer-range cycles of prosperity and depression—these had to be endured because of the belief that the system always adjusted itself in the direction of greater productive efficiency. The enonomists and businessmen generally agreed that creating a bigger economic pie for society was inevitably purchased at the expense of equality and security.

Not that there were no dissenters. Socialists—Marxian, Christian, or other—had argued for the need or the inevitability of a more equal distribution of wealth on moral, humane, religious, or "historical" grounds. But socialists were not the only foes of the market and its harsh disciplines. Fascists and traditional rulers sought to bend the forces of the market to uses of the state. Powerful businessmen had attempted by means fair and foul to carve out niches of security for their enterprises, far from the relentless pressure of an incorruptible marketplace. But to economists of the "central tradition," such ideas and activities were all wrong; they necessarily impaired the all-important goals of efficient production and economic growth, needed to overcome scarcity.

This, according to Galbraith, was the conventional wisdom which endured in the United States until the Depression. Under capitalism, economic life was highly insecure for individuals, businesses, and whole societies; rags and riches were doled out, but not indiscriminately. The economically virtuous were rewarded. The Depression brought into question the notion that economic insecurity had to be endured, that only the unworthy would suffer. Keynesian doctrine taught that an economy would not necessarily regulate itself for maximum efficiency. Into the conventional wisdom was incorporated the belief that during a depression, governments should intervene to increase aggregate demand. Productivity was still the focus of concern, but it was no longer at odds with the desire for broad national security, both economic and military.

Postwar prosperity for a time defused inequality as an issue in American society. As living standards rose throughout the economy, it became harder even for social reformers to get exercised over the unequal distribution of wealth. The prospect of increasing one's own take under the present system looked better than ever. Almost everybody seemed to have a vested interest in keeping things the way they were.

The modern industrial economy had become more than able to supply the needs of its population, said Galbraith. Yet despite the disappearance of scarcity, increased production remained the society's compulsive economic goal. The reason for this, Galbraith held, was twofold. First, the myth of scarcity still held men in thrall. Second, production had emerged as a means for increasing security and minimizing social conflict. A constantly expanding economy, depression-free, was the surest road to social tranquillity. There was an additional consideration. No more objective way of judging a society's progress could exist than data showing an increase in gross national product. Here was the anchor for economic analysis and planning. Galbraith warned that to cast doubt on the importance of production would be to "bring into question the foundation of the entire edifice."[47] But the time had come to test social action by new criteria, which would necessarily be difficult and subjective.

This was what Galbraith now proposed to do. An attack on the primacy of production emerged as the central purpose of *The Affluent Society*. Galbraith mounted an attack against what may be the holiest of capitalism's economic beliefs: the doctrine of consumer sovereignty. What consumers require will determine the nature of further production:

> If the individual's wants are to be urgent they must be original with himself. They cannot be urgent if they must be contrived for him. And above all they must not be contrived by the process of production by which they are satisfied. For this means that the whole case for the urgency of production, based on the urgency of wants, falls to the ground. One cannot defend production as satisfying wants if that production creates the wants.[48]

And in the United States, production *was* creating the wants. Advertising and salesmanship, Galbraith's old bugaboos, had a role which went far beyond substituting for price competition in oligopolistic and monopolistically competitive markets. They were responsible for convincing Americans of the urgency of hitherto unsuspected needs. These needs then became real. Such dependence of wants on the process of their satisfaction Galbraith christened the "Dependence Effect." For a substantial segment of the economy, increases in production did not therefore imply greater welfare, but merely a greater degree of want *creation* requiring a greater degree of want *satisfaction*.

Commitment to increased production was based on the myth of scarcity and the desire for economic security—harmless enough in themselves. But the reality of a constantly expanding American economy entailed serious consequences. Galbraith isolated two areas of the most serious concern. The first was inflation. In *The Affluent Society* he abandoned the idea that inflationary pressure was maintained by a high level of military expenditure. Rather, the culprit was the drive to increase production for its own sake. The society was "impelled by present attitudes and goals to seek to operate the economy at capacity where . . . inflation must be regarded not as abnormal but as a normal prospect."[49] One of the goals was the humane one of keeping down unemployment.

Still, the ill effects of inflation should not be minimized. The brunt of the burden of inflation fell on the weakest members of society, those with fixed incomes and those too poor or powerless to bargain for compensatory wage increases. This was destructive of what Galbraith called "social balance." Social balance was even more seriously impaired by the production ethic's discrimination between private and public goods. In a society where private production was more than adequate, the traditional bias against government spending remained strong. The apparatus of consumer marketing encouraged citizens to purchase all sorts of private goods, but no such apparatus encouraged expenditure for public programs and services. Here costs were always to be kept at a minimum. As a result, Galbraith argued, Americans were rich in private goods but poor in public ones. This imbalance was reflected in squalid cities, inadequate mass transit, and above all, in the continued existence of poverty in the wealthiest nation on earth. Moreover, the society's just claim to a healthy and attractive environment was constantly being sacrificed to industrial expansion, waste, and the frenzied and costly sales pitch.

The Affluent Society was a call to action. Improved social services, including health and housing programs, were needed to improve the lot of the poor and help them break out of the poverty cycle. In the society at large, goods could no longer be identified with happiness; attention had to be paid to the general quality of life. The cornerstone of Galbraith's program was education. Here lay his highest hopes. Education was the prerequisite for entrance into the growing body of white-collar workers, the "New Class." Characteristically, members of the New Class sought "exemption from manual toil; escape from boredom and confining and severe routine; the chance to spend one's life in clean and physically comfortable surroundings; and some opportunity for applying one's thoughts to the day's work."[50] Galbraith approved of these goals, but he resisted the temptation of equating their achievement with human happiness. The point was, however, that education provided a way to break the harmful and unnecessary dependence on production. More sophisticated, more guided by nonmaterial goals, the educated individual was less susceptible to having wants created for him by advertising campaigns. Likewise, an educated society was more able to determine where its well-being lay and better equipped to move in the desired direction. Education was thus not an expense but a crucial investment in the future of the country.

Galbraith strongly resisted the accusation that he was substituting his own desires for what society had, wittingly or not, chosen for itself:

> The question of happiness and what adds to it has been evaded, for indeed only mathematicians and participants on radio quiz shows are required to solve problems that can as sensibly be sidestepped. Instead, the present argument has

been directed to seeing how extensively our present preoccupations, most of all that with the production of goods, are compelled by tradition and by myth. Released from these compulsions we become free for the first time to survey our other opportunities. These at least have a plausible relation to happiness. But it will remain with the reader, and ultimately one hopes with the democratic process, to reconcile these opportunities with his own sense of what makes life better.[51]

Nonetheless, the message was clear. Poverty had to be eliminated, and the social balance between private and public goods had to be redressed. Expenditure on education must be increased, and standards of health and aesthetics established for the environment. These were real needs; expanding production was a mythical one.

The Affluent Society was Galbraith's most successful and influential book. Perhaps, according to the author's own portrayal of such causes and effects, it owed its success to the prior action of circumstance. One might question, as some economists did, the analysis of want creation and of the superfluity of increased production, but few could dismiss the list of America's problems and needs. The consumer society had run wild; salesmanship had infected the country. Social services were bad and getting worse. Poverty remained. More teachers and schools were required in order to accommodate the great postwar baby boom. The air stank from pollution.

Galbraith had crystallized much of what was wrong with the United States. Happily, the problem lay not in the nature of "the system" but in the nature of myth. Fears about the coercive power of big business were largely unjustified; of greater concern was its power over men's minds. Commitment to expanded production and the concomitant phenomenon of want creation had clouded society's vision. America had to wake up and take itself in hand. So, indeed, did the economics profession.

Surprisingly enough, both responded. The domestic poli-

cies of the Kennedy and Johnson administrations embodied
major elements of what could be called a Galbraithian program. Massive aid to education, the War on Poverty, support
for housing and urban renewal, consumer protection legislation, highway beautification, environmental protection, and
the pursuit of social justice for all citizens—these reflected a
new conviction that it was both necessary and possible to
improve the quality of national life. The ecology movement
exemplified and provoked a growing awareness of the destructive effects of unrestricted industrial expansion. In the
universities the new fields of urban, welfare, and environmental economics sprang up; "the limits to growth" became a
favorite topic for seminars and scholarly colloquia. Galbraith
could only consider himself vindicated if, fifteen years later,
much of *The Affluent Society* had become old hat.

VII

WITH *The Affluent Society*, Galbraith emerged as a full-
fledged pundit on the national scene. His thoughts were so-
licited by the press, and his presence by party givers. From
his typewriter flowed a stream of various prose pieces: journals, essays (serious, humorous, and autobiographical), lectures, political tracts, even fiction and art history.[52] Throughout, Galbraith remained a proponent of higher education,
aesthetic values, social and economic justice, and peaceful
relations with the Communists. He maintained his opposition
to ideological rigidity, militarism, and the production mentality. The best way to answer the Soviet challenge was by example; the strongest case for the American way would be
made by the removal of ugliness and inequity in our society
and by intellectual and artistic achievement. The image of the
United States was hurt by support of rascals and tyrants

abroad. It was madness to think of improving the world, or America's position in it, by nuclear holocaust.

Galbraith's preoccupation with politics and political power led him to write speeches for Adlai Stevenson in the 1952 and 1956 presidential campaigns. In 1960 his political ship came in. As high-level adviser and campaigner for his old acquaintance and one time student John Fitzgerald Kennedy, Galbraith was in the thick of the action when the new administration came to power. What position, if any, he would assume with Kennedy in office was, however, highly unclear. As a less than orthodox economist, Galbraith was deemed unsuitable for membership in the President's Council of Economic Advisers. He was considered too liberal and too outspoken for a cabinet position. A diplomatic post seemed more in order. While waiting for the word, Galbraith meditated on the possibility of his filling President Kennedy's unexpired term in the U.S. Senate. Mr. Kennedy scotched this idea, urging on him instead the ambassadorship to India. A leave of absence from Harvard was obtained, and in due course Professor Galbraith received Senate approval in the diplomatic post.

Galbraith served with distinction in India for two and a half years in a job which turned out to entail far more than ceremony and public relations. India's takeover of Goa, its border war with China, and the crisis with Pakistan over Kashmir occurred during Galbraith's term from 1961 to 1963. Throughout his term the Harvard economist traveled widely, spoke with all sorts of people, took off his shoes and waded in rice paddies, and demonstrated insight and considerable diplomatic skill in helping to restore peace and safeguard American interests. Galbraith's common sense is sometimes masked by his high style and elegant irony, yet his success in India stands out in the generally bleak picture of United States foreign policy during the 1960s. In no subsequent period have American relations with India been so good.

John Kenneth Galbraith

As ambassador, Galbraith found time to pursue an old interest in economic development. In 1962 he brought out *Economic Development in Perspective,* a slender volume of five lectures delivered the year before at various Indian universities. He accorded high priority for developing nations to the claims of social justice and education, and he urged that long-range public planning be combined with short-term private decision making.

Galbraith also acquired a taste for classical Indian art and a reputation for squiring glamorous women around, notably Jacqueline Kennedy and the actress Angie Dickinson. He wrote a pseudonymous satire, *The McLandress Dimension,* which seemed to burlesque his own preoccupation with self. But this was all, ultimately, peripheral. He was growing tired of working for the State Department and was anxious to get back to his real work. In the summer of 1963 he resigned, returning to Harvard for the fall semester.

Six years later, when he published a slightly expurgated version of the journal he kept while in India, Galbraith described the Kennedy years as "a gay and rewarding interlude in our history."⁵³ The interlude was especially exciting for the academics whom the president brought to Washington. The charm, sophistication, and intelligence of Jack Kennedy affected them all, and Galbraith not the least. If anything could be done, this debonair President could do it. Then came the tragedy in Dallas, and the style of the American presidency changed. Gradually, many of the leading academics departed from Lyndon Johnson's administration, conscious of their diminished role.

Galbraith, who had already departed before LBJ took over, handled himself gracefully in the succeeding years. He supported the Great Society domestic programs of the Johnson administration without allowing himself to be drawn into apologetics for the Vietnam war. He became an active member of the Americans for Democratic Action, serving as chairman from 1967 to 1969. In 1968 he joined the movement to "dump Johnson," securing ADA support for Senator Eugene McCarthy in the race for the Democratic presidential nomination. Now and again, visions of becoming senator from Massachusetts danced before Galbraith's eyes, but this seemed fated not to be. He had gone as far as he would into the fires of politics, and had emerged with principles unsinged.

VIII

THE NEW INDUSTRIAL STATE appeared in 1967. Galbraith reminded his readers that "this book had its origins alongside *The Affluent Society*. It stands in relation to that book as a house to a window. This is the structure; the earlier book allowed the first glimpse inside."[54] *The Affluent Society* had spotlighted the unfortunate effects of the American industrial system; these, Galbraith had argued, could be alleviated once outmoded beliefs were set aside. *The New Industrial State* sought a deeper understanding of how the system functioned. Like *American Capitalism*, it aimed at presenting a gestalt of economic life; the industrial state was a coherent, interlocking whole with specific institutional and psychological contours. Galbraith contended that "imperatives of technology and organization" made for a broad convergence of industrial systems, whether capitalist or socialist.[55] The images of ideology had thus become obsolete as descriptions of such systems. The industrial state created severe problems for itself; at the same time, however, it contained the seeds of its own salvation.

Galbraith built his new structure around two facets of *The Affluent Society*: the New Class and the quest for economic security. Technology was defined as "the systematic application of scientific or other organized knowledge to practical tasks."[56] This was the hallmark of modern industrial production. The most pressing demands it placed on society were for skilled manpower and for the security to plan ahead. The one eliminated the preeminence of the entrepreneur, the other the preeminence of the marketplace.

Security in the conduct of one's economic affairs was an age-old desire. Under conditions of high technology it became a pressing need. A firm's day-to-day operation was so complex that small changes in its costs, in the nature of the materials it used, and in the structure of demand could cause major disruption. To get a new good (or a new wrinkle in an old one) into production could take years of development and millions of dollars. Throughout, an extremely solid basis was needed for decision making, for

> in addition to deciding what the consumer will want and will pay, the firm must take every feasible step to see that what it decides to produce is wanted by the consumer at a remunerative price. And it must see that the labor, materials and equipment that it needs will be available at a cost consistent with the price it will receive. It must exercise control over what is sold. It must exercise control over what is supplied. It must replace the market with planning.[57]

The higher the technology, the more the market was replaced. This happened in various ways. One was vertical integration, whereby a firm manages several levels of production in the transformation of raw material into a final good. This process "internalizes" the market within the firm. The market could also be controlled by sellers, through advertising or by sheer market power. Finally, long-term contracts could be used to suspend operation of the market.

Advertisement and salesmanship—the management and creation of demand—thus became for Galbraith part of an overall strategy by which industry gained security for planning and operation. This strategy was necessitated by the use

[137]

of sophisticated technology, and it depended above all on size and power. The enemy of the market was not ideology but technology, or more precisely, the *people who controlled technology*.

The amount of knowledge—scientific, technical, managerial, economic, legal—required to run modern industry far exceeded the capacities of individual entrepreneurs. Responsibility for a firm's operation had increasingly come to rest in the hands of the skilled researchers, technicians, and executives who possessed that knowledge. The success of the firm depended on the individual talents of these operatives and, most importantly, on their ability to work together as a unit. They were the children of the engineers described by Thorstein Veblen in *The Engineers and the Price System*.[58] Together they made up what Galbraith called the "technostructure."

Economists had traditionally spoken of three factors of production: land, capital, and labor. Prior to the Industrial Revolution, according to Galbraith, land was the predominant factor. He who possessed land could attract—and control—both capital and labor. With the industrial revolution, however, power shifted to capital. The capitalist, not the landowner, was in the driver's seat. Later, with the rise of unions, substantial power also accrued to labor. Now, Galbraith held, a new factor of production had emerged in the person of the technocrat. Moreover, this new factor had gained predominance:

> Power has, in fact, passed to what anyone in search of novelty might be justified in calling a new factor of production. This is the association of men of diverse experience or other talent which modern industrial technology and planning require. . . . It is on the effectiveness of this organization, as most business doctrine now implicitly agrees, that the success of the modern business enterprise now depends.[59]

[138]

The crucial fact about the technostructure was that it ran the means of production but did not own them. Divorced from ownership, the technostructure did not itself receive the profits of its work; these went, at least in the United States, to stockholders. For this reason, Galbraith argued, industry could no longer be assumed to operate according to a strict rule of profit maximization.

> Loss can destroy the technostructure; high revenues accrue to others. If, as will often happen, the maximization of revenues invites increased risk of loss, then the technostructure, as a matter of elementary interest, should forgo it.[60]

The prime goal of the technostructure thus became its own security. This was achieved in a number of ways. A firm had to be operated at a level of success which assured its continued existence and the continued autonomy of the technostructure. The stockholders had to be kept happy, to prevent their interference with the smooth functioning of the technostructure. Commitment to technological progress also served the interests of the technostructure. Most important of all, however, was industrial expansion. This provided the greatest prospect for a secure and expanding technostructure. The increased production of goods was the critical concern.

Galbraith had reached what he considered to be a more satisfactory explanation of the concept of dependence on production than he had advanced in *The Affluent Society*. Circumstance, Galbraith believed, determined human activity to a greater extent than ideas. By emphasizing the importance of economic myth, the earlier book had tended to minimize the effect of an underlying institutional reality in promoting the drive for more goods. This drive remained unimpaired despite the social programs of the Kennedy and Johnson administrations and the growing awareness of the need to restore social balance. The explanation lay in the requirements of technology and the technostructure. In the new industrial state, highest priority was given to the goal of economic security, and this meant continuous expansion of production.

The results were awful. In the mindless pursuit of un-hindered growth, environmental standards of health, safety, and aesthetics were violated. Though serving no one's pur-pose, inflation and the wage-price spiral were unavoidable side-effects of the new industrial state. An imbalance between public and private goods appeared inevitable. Yet Galbraith did not feel that all was lost; again, there was hope in educa-tion.

To perpetuate itself, the technostructure had an ines-capable need for higher education. The skills and knowledge required to run modern industry and keep the system expand-ing demanded advanced academic training. But this posed a real dilemma for the educators, scientists, technologists, economists, and other members of the intelligentsia: they could either supinely serve the needs of the industrial system, or they could "strongly assert the values and goals of edu-cated men—those that serve not the production of goods and associated planning but the intellectual and artistic develop-ment of man." Galbraith found it "hard to believe that there is a choice."[61]

But were the educators strong enough to impose their val-ues and goals on the industrial system? Given the system's critical need for educated and trained manpower, Galbraith suggested, the seemingly weak educational establishment had more power than it realized, if it only had the wit and deter-mination to use it. The goals of the mature corporation would be "a reflection of the goals of the members of the techno-structure."[62] If education broadened the understanding, heightened the aesthetic sense, and ennobled the social pur-pose of the technostructure, the corporation could become an instrument for social change and national betterment. It would mobilize talent to fill in the nation's "planning lacunae," such as urban and interurban surface transit, real estate development, and urban and suburban housing. Under the influence of education and the pressures of the political process, the new industrial state would accord precedence to environmental concerns and to the need for better social balance. Here was the hope for "salvation":

> The industrial system, in contrast with its economic antecedents, is intellectually demanding. It brings into existence, to serve its intellectual and scientific needs, the community that, hopefully, will reject its monopoly of social purpose.[63]

The New Industrial State thus concluded with a call to arms to the New Class—the class of intellectuals and professions that he believed was becoming the dominant one in American society. The goals of the nation, the public purpose, were its to decide. For his own part, Galbraith made no secret of how he wanted the decisions determined. His values were on the table, and he seemed reasonably confident that other intelligent people would adopt similar values.

Five years later, when he published *Economics and the Public Purpose*, Galbraith was much less confident about the intellectuals, including the economists. Vietnam had marked the downfall of "the best and the brightest," the Robert McNamaras and McGeorge Bundys and Walt Rostows. The Nixon administration, in ending the draft, had stilled the revolt on the campuses and had infected the society with its own cynicism. Persistent inflation—and the unwillingness of the economists to recognize that its root causes lay in the structure of the modern economy, or to accept the necessity of controlling it directly—indicated to Galbraith the bankruptcy and rigidity of the current economic establishment, dominated by the disciples of John Maynard Keynes and champions of the "neoclassical synthesis" of Paul Samuelson.

In *Economics and the Public Purpose,* which he regarded as the last part of the trilogy that included *The Affluent So-*

ciety and *The New Industrial State*, but which also had "some genes, though not many, from yet another volume—*American Capitalism*,"[64] Galbraith boldly set out to move both the economics profession and the literate public beyond Keynes and Samuelson. The most serious deficiency of "mainstream" economics, in his view, was that it had no useful handle for grasping the most critical problems that beset modern society: enormously costly preparations for war that distort the use of national resources and lead to trigger-happy behavior; extreme economic inequality, which leads to extreme social tensions; and the uglification of the civic and natural environment, which leads to the debasement of man, woman, and child.

By slurring over the role of power in the economy, he stressed, mainstream economics, with its model of a self-adjusting price mechanism, had destroyed its tie to the real world and had provided fancy apologetics or obfuscation for the not particularly invisible hand of the great corporations. And by exposing the reality of that power and the inadequate or poor social performance of even that part of the economy that is more or less competitive, Galbraith hoped to help his fellow economists and other readers to emancipate themselves from the follies of current conomic belief and contribute to the emancipation of the state from the dominance of what he now called the "planning system," essentially the large corporations and their political and bureaucratic allies and agents in government.

Galbraith had now gone beyond his earlier books, with their focus on the large corporations, to describe the whole modern capitalist economy, which he saw as roughly split between the planning system and what he called the "market system," a collection of relatively small but still imperfect competitors and partial monopolists that included such producers as farmers, television repairmen, retailers, small manufacturers, medical practitioners, photographers, and pornographers. He saw the market system as the exploited and relatively feeble half of the economy, a view that went back to his earliest days as a graduate student at the University of California. But the market system in the United States obviously includes some not particularly exploited people, such as millionaire doctors, real-estate operators, small manufacturers, entertainers, and indeed even some big-time farmers and pornographers.

Galbraith continued to insist that although some members of the market system have made out quite nicely, it is the planning system that dominates the state, unbalances social and economic development, exacerbates inequality, corrupts foreign policy, and befouls the atmosphere. So Galbraith, after his long flirtation with the establishment and his tour of the corridors of power, was at last proposing to change the system radically, though gradually. He called for a "new socialism," one that would extend to the market system the organization, protection, and power of the planning system. This was, in fact, where he believed the system had already been going, in such quasi-market areas as agriculture, health services, housing, and transportation. Socialism might emerge from the womb of competitive-business failure.

His new brand of socialism would include the establishment of guilds of small retailers, repairmen, domestics (no longer servants), auto-body shops, and whatnot; the antitrust laws, he said, pursuing his old bias against them, would be suspended so that these small businessmen could organize themselves, mobilize their power, and fix their price structures, just as the great corporations do. He proposed to provide a net to stop even the lowliest members of society from crashing to the ground; everyone in Galbraith's new socialist society would have the assurance of a decent annual income, set to provide a bit less than one could earn by working within the planning system. The tax system would be made far more progressive, and this, he said, would make the economy far more stable. Galbraith's new socialism obviously, and not coincidentally, had much in common with Senator George McGovern's 1972 campaign program. Of course, the Galbraithian manifesto was more lucid, witty, and persuasive —partly because it was unencumbered by confusing or alarming arithmetic. He recognized that his own program, including proposals for a redistribution of income, might not yet sell, even after Watergate. But he insisted that in the end it would prevail, because the underlying forces of economic history and of the "public cognizance"—recognition by the public of its own best interests—would so ordain.

Galbraith had at last come out as a socialist. This promised to be publicly useful, since it would add clarity and realism to the debate—first among the economists, whom Galbraith taunted as acolytes of the neoclassical myth of the market, and later among the politicians and the voters. But whether the economists would play an important part in helping the public to understand and deal with "the gravest problems of our time"—producer dominance of the society, unequal social development and income distribution, the wasted environment and victimized consumer, inflation and the need for both national and international planning—was up to the economists themselves:

> They can, if they are determined, be unimportant; they can, if they prefer a comfortable home life and regular hours, continue to make a living out of the infinitely interesting gadgetry of disguise. They will, as was the case in the summer of 1971, when price controls were imposed in the United States, or a year later when they were put into effect in Britain, have few or no words of guidance or advice on great issues. They will be socially more irrelevant than Keynes's dentist, for he would feel obliged to have a recommendation were everyone's teeth, in conflict with all expectation, suddenly to fall out.

> Or economists can enlarge their system. They can have it embrace, in all its diverse manifestations, the power they now

disguise. In this case, as we have seen, the problems of the
world will be part of their system. Their domestic life will be
less passive. There may be a contentious reaction from those
whose power is now revealed and examined. And similarly
from those who have found more comfort than they knew in
the fact that economists teach and discuss the wrong problems
or none at all.[65]

Galbraith has continued to try to force debate on the central issues of socialism, freedom, corporate power, and the
role of the market. He still feels, as he has felt throughout his
career, that it is a waste of time, energy, and legal fees to try
to break up corporate power by fractionating business enterprises. The real issue, he holds, is not how to diffuse power
but who will control that concentrated power, the corporate
barons or the public. The concentration of power—especially
in the light of the Nixon presidency, which exploited the apparatus of both government and business—worries Galbraith, but he has continued to maintain that "the public
cognizance" and a strengthened Congress, "the natural voice
of the public purpose," would protect the nation from au-
thoritarianism.

I X

YET the bulk of American economists still believe it is a
good idea to diffuse power, and that the market remains a
useful instrument of diffusion. They are afraid that Galbraith's new socialism could turn out to be as bureaucratic
and corrupt or corruptible as the existing system, perhaps
even more so. They have refused to be bullied or shamed by
Galbraith into accepting his economic philosophy, his concept of the very wide scope of economics, or his methodology. The quantity of undocumented assertion in Galbraith is

too large for a profession that prides itself on its scientific qualities. The profession was charitably willing to ignore the lack of substantiation in Galbraith's best-read book, *The Affluent Society*, because so much that was so evident in everyday life sustained his argument. But statements about the nature of industrial organization were a different matter. Who knew whether the profit motive still reigned in this or that great corporation, let alone throughout the structure of modern industry? Who could even define operationally what power, whether economic power or some other kind, actually is? How could one assess the validity of Galbraith's assertions about the impact of technology or technological change upon the society? Readers, lay and professional, were being asked to take a great deal on faith. Economists, each displaying his superior knowledge of his specialty, found much to fault in Galbraith's work. He has received, and doubtless provoked, a good deal of disagreeable sniping.

His work has also called forth charges of elitism. Economists have accused him of seeking to substitute his own values for those held, for good or ill, by the rest of the society; this was antidemocratic of him, or arrogant. Such has remained the most common criticism of Galbraith's thought. It is also the most unfair. To accept it is to miss the point of all his work.

Since he began his work in the 1930s, Galbraith's prime concern has been with the economic power wielded in the industrial sector of the economy. Monopoly power produced price rigidities, exacerbating the Great Depression and imposing particular hardships on farmers and other groups incapable of protecting their own prices or wages. "Countervailing power" helped to stabilize and equalize power in the system but made the economy extremely inflation-prone. Throughout, Galbraith has urged a significant role for government—to check inflation, to restore balance between private and public goods. In a democracy, government reflects the will of the people, and government must act in the public interest to offset the power of the industrial system. What prevented this was a tradition of economic thought and prejudice that said that economic power did not exist (or could not be identified) and that governmental intervention would only make things worse and would surely make the economy less efficient.

Classical economic doctrine had provided an invisible hand of the market to ensure maximum efficiency in the allocation of resources, but Galbraith held that the invisible hand had atrophied. Economic power had irremediably supplanted the free market in the modern industrial state. The only question was who would wield power and to what purpose. Galbraith's preference was for citizens to do so through the agency of a government informed by deep, sensitive, and far-sighted economic thinking. This was the purport of his presidential address to the American Economic Association:

> Economics does not become a part of political science. But politics does—and must—become a part of economics. . . . If we accept the reality of power as part of our system, we have years of useful work ahead of us. And since we will be in touch with real issues, and since issues that are real inspire passion, our life will, again, be pleasantly contentious, perhaps even usefully dangerous.[66]

Galbraith has had little to contribute to the analysis of technical economic problems, little to add to the ongoing refinement of economic theory. He has, however, always remained in touch with real issues. This is why his books have sold so well. At their best, they have pointed up areas of concern just below the level of public consciousness, and have challenged traditional and stale ways of thinking. Galbraith has exposed some of the capitalist system's failings brilliantly and sometimes even hilariously.

Nevertheless, his style, flavor, and scope still provoke

heated controversy among economists. At a recent annual meeting of the American Economic Association, the following exchange occurred:

"Is Galbraith really an economist?"

"He happens to be the only economist."

There certainly has been a grand and glorious unity to his oeuvre, running from his first work on honey and depressed prices in the California produce market; through his battles with businessmen as a government wartime price controller and his assaults on the Pentagon and on his fellow economists as the self-deceived front men for big business and as innocents in a world of power brokers and power wielders; to his advocacy of a transformed American society in which power would revert from the industrial elite to the people, and to their educated and presumably socially responsible leaders. His is indeed the voice of a new socialism, of a happier and healthier polity and economy, both to be attained by peaceful and painless democratic methods. Galbraith's works have been manifestos without parallel in the history of political reform or revolution—funny and literate works that heal as they wound. But is this grace and good humor sufficient? Does the humorist ever change anything?

Wassily Leontief

Apostle of Planning

"To predict the broad developments of the future is no more of a *tour de force* than to divine those of the past. . . . If past events have left their traces, it is reasonable to imagine that those still to come have their roots."
                    —*Honoré de Balzac*

I

WASSILY LEONTIEF has devoted most of his life to technical economics, and particularly to the development of the economic model he conceived in his early youth: input-output analysis. Except for those economists who have worked directly with him, he has few disciples; it is hard for most students to be enthralled by empirical data laid out on a matrix —as T. S. Eliot might say, "like a patient etherized upon a table." Yet in the midst of his technical labors, Leontief has dreamed of creating a tool that societies could use to achieve a better life for all, an instrument that would give humanity greater control over its destiny.

This does not imply that Leontief, born in Russia before the revolution, is a secret Bolshevik. On the contrary, he left the Soviet Union after the revolution because he was a rebel against authority. As an economist he has sought to find a way of reconciling planning with political liberty. That was a dream that, for many, burned most brightly during 1917 and the years immediately following. For a time the dream seemed to have been snuffed out by the reality of life under the Soviet dictatorship.

Yet Leontief, after having spent virtually his entire adult life in the United States, has returned to that dream of democratic economic planning in the interests of a healthier society. His technical labors on his input-output mechanism won him the Nobel Prize in Economic Science—and emboldened him to carry his message into politics in his adopted country.

Leontief's forebears were peasants. His great grandfather managed to leave the land—"it was not like America; our peasants were still in virtual slavery until late in the nineteenth century," says Leontief—and made it to St. Petersburg, where he became a merchant. Leontief's paternal grandfather started a textile mill, "a classic development." There is a branch of the Leontiefs in England, named Cheshire—"they married English girls."[1]

The Russian Leontiefs grew rich. "Grandfather built a great house, a beautiful old-fashioned Tolstoyan mansion," recalls Leontief. "I have visited it lately . . . the ballroom is subdivided into small apartments." Leontief's own father then made "the normal transition from the entrepreneurial class to the intellectual class," becoming a professor of labor economics at the University of St. Petersburg.

Wassily Leontief vividly remembers the years of the Russian Revolution—"the country plunged into deep mourning the day of Leo Tolstoy's death; stray bullets whistling by during the first days of the February Revolution; and Lenin addressing a mass meeting from a high tribune in front of the Winter Palace." He was graduated from high school in 1921 at the age of 14 and entered the University of Leningrad at 15, where he had a turbulent academic career. He studied philosophy, sociology, and finally economics, and was awarded the degree of "Learned Economist" in 1925 before his 19th birthday; but he kept getting into trouble over his outspoken criticism of the regime. "I was pretty free-speaking then as now," he says, "and was put in jail from time to time." But he claims he maintained "good relations with the Communists," though he himself was "a noncommunist socialist, a Menshevik." Why did the Russians keep letting him out of jail? "They thought they could reeducate me. . . . The last time they let me out was during the New Economic Policy. . . . You see the scar on my left cheek. [It is a deep one.] I had a tumor that was diagnosed as malignant, so they let me out. The doctors used a rib to replace my jawbone. But the growth proved not to be malignant."

After he got his degree as "Learned Economist," he was offered a job as instructor in economic geography ("a neutral subject"), but within six months applied for and received a passport. He left to study in Berlin, but the Germans did not recognize his Russian degree; so he had to go back to school to prepare for and pass examinations in Greek and Latin in order to enroll at the University of Berlin. He stayed on in Berlin until 1928, where he served as an assistant to the great German economic historian Werner Sombart and studied mathematical statistics under Professor Ladislav Botkevich. Leontief's family did not leave Leningrad with him at first; he supported himself in Berlin by writing for trade journals. In 1927 his father, who had become a bureaucrat, was sent to Berlin as a representative of the Soviet Ministry of Finance, bringing his wife out with him. He "just escaped with his skin," says Leontief, because when the Stalinist purge trials began, his father was accused of "all sorts of things," especially joining antigovernment conspiracies.

I I

LEONTIEF appears to have hit very early upon the basic idea for input-output economics, for which he was to receive the Nobel Prize in 1973. In 1925, in Berlin, when he was just 19 years old, he published an article, "Die Balanz der Russischen Volkwirtschaft—Eine methodologische Untersuchung" ("The Balance of the Russian Economy—A Methodological Investigation") in a German economics journal.[2] The article was translated into Russian and published two months later (without Leontief's permission or knowledge) in the principal Soviet planning magazine; that Russian translation of the German article—Leontief's first—is the one the Russians always refer to in establishing their claim to have founded input-output economics. This enabled them to accept input-output as a native Russian product, not a bourgeois import.

In 1927, Leontief joined the staff of the Institute for World Economics at the University of Kiel, where he worked on statistical methods of deriving supply and demand curves, the same subject on which Milton Friedman was later to work under Professor Henry Schultz at the University of Chicago. Leontief's Ph.D. thesis was titled "Die Wirtschaft als Kreislauf" ("The Economy as a Circular Flow"). He describes his work as an effort to produce "a dynamic general equilibrium model"—that is, a mathematical description of the way the economic system moves toward equilibrium over a period of time. In this study, says Leontief, "I discovered the business cycle for myself—because the dynamic model had lags built into it, and the level of economic activity went up and down because of those lags." His whole bent was to link elements of the economy together. He was searching for a middle course between the empirical analysis of supply and demand for particular products and the abstract construction of systems meant to represent the economy as a whole.

His last major job in Kiel, he remembers, was a study of the steel industry, in which he worked with real data to derive the supply and demand curves not only for semifinished and finished steel products but also for ore, coke, and the labor going into steel and its raw materials. "I was starting to pull together a big molecule," he recalls, "out of those separate atoms. I was trying to show how the supply and demand for steel depended partly on the supply and demand for all those other things—and that it involved technological transformations." But he decided that it would be too difficult to try to put together a model of the entire economy on the basis of separate supply and demand curves for every product. What he needed was to figure out a way to map the flow of production from one sector of the economy to another.

In thinking how to go about that task, he says, "the classical economists certainly influenced me. It was the Physiocrats. It was Sismondi. And it was certainly Marx as well." Leontief was steeped in the classics. "As a young student in Russia," he says, "I spent long days and evenings in the public library in Leningrad, which was magnificent—nearly as good as the Bibliothèque Nationale in France. I read all the economics books from beginning to end—in French, in English, in German. I still think classical economics is a magnificent structure." From his classical background, he believes, the theoretical framework of input-out "naturally suggested itself."

The most important of the classical progenitors of input-output was undoubtedly François Quesnay, physician to Louis XV and Madame de Pompadour. Quesnay turned to economics in his later years. He drew up the *Tableau économique*, describing the flow of production and expenditures among farmers, manufacturers, and landowners. This *tableau* showed how the output of the farmers (the "productive class," as the Physiocrats saw them), was absorbed by the landlords and the "sterile class"—the manufacturers. In exchange, the farmers obtained some of the output of the manufacturers, and at the same time received the means of paying their rent to the landlords. Both the landlords and manufacturers, on the next round, bought more food from the farmers with the money they had received; so the circular flow went on and on.

Quesnay's primary *tableau* depicted a society in equilibrium, but he developed his model to show that if the landlords increased their expenditures for manufactured goods and reduced those for food, the effect would spread throughout the economy, first raising the income of the "sterile class" but eventually throwing the economic system out of balance. The landlords' income ultimately depended on the farmers' "net product," which declined as the landlords shifted their trade to the manufacturers. Since the income of both the farmers and the landlords would then be lower, the manufacturers would also suffer from a lack of customers.[3] The point of Quesnay's analysis was to prove the Physiocratic doctrine that agriculture was the source of all wealth. Quesnay's *tableau* model, based on both observed and conjectural data, had excited great wonder and admiration in the eighteenth century. But for Leontief it was a primitive didactic toy. He wanted to build an economics that could deal with the world as it really was.

III

AT the Institute for World Economics in Kiel, the staff would come in at eight o'clock in the morning and work until noon, take three hours off for lunch, and then go back to their desks from three o'clock to seven or later. One day, during the leisurely three-hour lunch break, Leontief and his colleagues were in a coffeehouse, drinking "nice Kiel coffee," seated next to a group of Chinese, "real gentlemen," he says, not ordinary workers. A conversation developed between the Kiel economists and the Chinese strangers; a month later, Leontief received a telephone call at the Kiel Institute from the Chinese ambassador in Berlin, asking whether he would be willing to go to Nanking as an economic adviser in the Ministry of Railroads. Leontief gladly accepted a one-year contract.

At the end of 1928, he booked first-class passage from Marseilles to Shanghai on what he calls "a slow boat to China." "It was my introduction to the underdeveloped countries," he says. "I left the boat in Egypt, Arabia, Ceylon, . . . for brief visits. I went into the villages wherever we stopped

—it was my first taste of Asian poverty. It was a garbage can—it's still a garbage can."

In Nanking, Leontief had a variety of assignments. He negotiated with China's creditors, especially the Compagnie Generale de Credit of Belgium, who, he says, were in "a weak position, since their property could be expropriated. . . . Since we had no chance of getting any new loans from them, we were in a strong bargaining position." Leontief also worked on the planning of new rail lines. He had only the most scrappy data for planning purposes, so he decided to create his own. He persuaded his minister to get an airplane; he had cameras mounted on it and sent it out to photograph the crops. He used these pictures, on a sampling basis, to make estimates of farm production by regions as a basis for planning rail lines.

He remembers China as a happy experience. "I rode on trains that came under fire from bandits; I remember lying down on the floor. There was fighting in the north with the Russians. The Russian planes bombed the Chinese with watermelons. They threw the watermelons down on them. This was an idyllic world, you see."

When his year's contract was up, Leontief returned to Kiel and went back to his statistical studies at the Institute. His published work attracted attention in Britain and the United States, especially at the National Bureau of Economic Research, whose leaders were eager to shift American economics to a stronger empirical foundation. Wesley Mitchell, the president of the National Bureau, and Professor Edwin F. Gay of Harvard, one of the Bureau's founders and a director, wrote to Leontief, inviting him to join its staff. He accepted and sailed for New York in early 1931; the National Bureau staff man who met him at the dock was Simon S. Kuznets, who a few years later would bring Milton Friedman to the National Bureau.

Leontief stayed only briefly at the National Bureau in New

York; he says he did not like its narrow focus on index numbers, on efforts to trace the wiggles of the business cycle without an adequate theoretical foundation for analysis. "I organized a seminar," he says, "a subversive group, that was trying to smuggle theory into the place." Two months after arriving at the Bureau, Professor Gay invited him to give a lecture at Harvard. "He was a very important figure; it was a command performance," says Leontief. (Paul Samuelson, on arriving at Harvard, had refused to take Gay's course on the history of economic thought.) Apparently the lecture went well, for Leontief was offered a post at Harvard. He was eager to accept but said he could do so only if he could go on with some research he had just undertaken. It would involve detailed statistical computations and would require financial support. Leontief then described in a letter to Gay his plan for constructive tables to show the flow of goods among sectors of the American economy.

Gay replied that the Harvard economics department had considered Leontief's request, found it unpromising, but would nevertheless make him a grant of $1,400 so that he could employ a research assistant for one year. After that time, he should give the committee a report of his findings. Leontief accepted, and the work on input-output began in earnest. Leontief and his new wife, Estelle Marks, a poet, moved up to Cambridge.

IV

OVER THE YEARS, Leontief has not been a prolific writer, but he has built steadily, patiently, solidly. Before he wrote the 1936 paper that really introduced the input-output theory, "Quantitative Input and Output Relations in the Economic System of the United States,"[4] he had been working out his general equilibrium theory for years, and he had then spent his first four years at Harvard gathering information for the first input-output tables. It was another five years before he published his first book, *The Structure of the American Economy*, *1919–1929*,[5] which expanded on the original paper and further developed and explained his system. The long and tedious work of calculating the equations had to be done with hand calculators; it was the coming of the electronic computer that made input-output a usable tool for government and business.

Leontief has not been a one-idea man. He has turned his mind and writings to many other facets of economics and the human experience. The dominant theme that runs through his work is that economics should be an empirical and applied science. He is impatient with highfalutin model building, considering it no more than half the job. And he is also impatient with economists who start their theorizing from unspecific and faulty premises and who indulge in circular arguments. Although thoroughly trained and highly gifted in mathematics—indeed, he seems to think mathematically—he often writes skeptically of the application of higher mathematics to economics. He works hard and gives short shrift to economists unwilling to roll up their sleeves and do the dirty work of collecting and analyzing empirical data. He has turned his attention to a wide variety of technical and empirical problems: linear programming, problems of aggregation, automation, real quantities as contrasted with index numbers, economic growth, the economic impact of science and technology, disarmament, foreign trade, pollution of the environment, and the interrelation of political, social, and economic forces.

But input-output has certainly been the center of his work and the key to his attack on many other problems, including his efforts to solve global environmental disruption and facilitate world economic development. His Nobel Memorial

Lecture, entitled "Structure of the World Economy: Outline of a Simple Input-Output Formulation," described the world in terms of 28 groups of countries, with some 45 productive sectors; it set out to analyze environmental conditions in terms of 30 principal pollutants. The aim of this tremendous effort, which is being sponsored by the United Nations, is to help member states review the economic gains and environmental damage resulting from past economic development and to help them lay plans for reducing mass poverty and unemployment in ways that will at the same time preserve and even improve the global environment.

What is this all-purpose tool which Leontief thinks can be put to use for analyzing and planning the world economy? In essence, it is a way of seeing the economy as a system of simultaneous equations. While Quesnay and the Physiocrats of the eighteenth century provided the first inspiration for Leontief's work, it was the nineteenth-century French economist Léon Walras who first showed how an economic system could be set up mathematically. He developed a system of simultaneous equations to show the way the numerous markets in the economy are linked; Adam Smith and David Ricardo had done this in a literary way. Walras's equations showed how the different quantities of each of the productive services combined in turning out each product—and the total output of the entire system.

Such equations formed the basis for Leontief's system, but not without modification. The Italian mathematical economist and sociologist Vilfredo Pareto estimated that for every 100 persons exchanging 700 goods, 70,699 equations would be needed under Walras's system. For a world of 4 billion people, the number of equations would approach infinity, a problem no analyst could solve. Leontief and his followers have therefore simplified and substantially reduced Walras's system to a more feasible number of equations.

Leontief often describes the economic system as a gigantic

computing machine, tirelessly grinding out solutions to an unending stream of quantitative problems, such as the more efficient uses of labor, capital, and natural resources, or the proper balance between the production and consumption of goods, or the balances among consumption, savings, and investment. But like a computer, the economic system is subject to breakdowns, and Leontief thinks a blueprint of the economy's working parts is needed before the defective parts can be repaired.

Hence his input-output system, a guide or manual for the real economic machine. The flow of goods and services among the different sectors of the entire American economy —or that of any other country, group of countries, or the world as a whole—can be laid out on a giant ticktacktoe grid or matrix. All industries are listed down the left side of the grid, one industry to a row, and all industries are also listed across the top of the grid, one industry to a column. The figures in the cells of the horizontal rows represent the outputs (goods and services the industry turns out and sells to others) of the industry named at the left, and the figures in the cells of the vertical columns represent the input (goods and services the industry buys from others) to the industry at the top of each column.

Thus, the output of each industry (or sector) is always the input of some other industry, and each figure represents both an input and an output. In other words, each industry's sale (output) is another industry's purchase (input). The sum of all the outputs of an industry represents the total sales of its products. The sum of all the inputs of an industry represents its total purchases from other industries, used in achieving its total production for the year.

Such a table can be used as a planning tool—for instance, to determine where shortages will occur in wartime. Leontief points out that if a usable input-output system had existed at the start of World War II, it would have been easier to predict where shortages would occur. President Roosevelt called for 50,000 airplanes, and almost anyone could have predicted that the country would have to produce more aluminum. But it was not so apparent that building aluminum potlines would be hampered by a shortage of copper (a problem that was finally overcome by borrowing silver from Fort Knox).

To make such problems more readily apparent—and to turn a given set of input-output relations drawn from empirical data into a tool for predicting or planning future needs—a table of *input-output coefficients* is drawn up. In a coefficient table, the figure in each cell expresses the ratio of the input from the industry in whose row the cell appears to the total output of the industry in whose column that cell appears.

Studying the relationships of interindustry transactions on an input-output table, one can see how much steel, aluminum, rubber, electrical energy, and other items the auto industry has to buy in order to produce the cars it turns out year after year. Taking a closer look, one can see how the auto industry's purchases of glass depend partly on the glass industry's demand for motor vehicles.

Drawing up such a set of interlocking tables for the entire economy is a massive and meticulous job. Even Leontief ruefully admits that gathering such a mass of information "has little appeal to the theoretical imagination."[6] Actual records of interindustry transactions have to be searched and compiled, then supplemented and refined. Employing one of his favorite metaphors, Leontief says that like individual atoms and molecules, the individual transactions are far too numerous to observe and describe in detail, so they must be classified and aggregated into groups.

The first input-output tables which Leontief and his assistants at Harvard prepared were based mainly on data from the U.S. Census of Manufactures for the years 1919 and

1929 and were broken down into only 42 sectors. They were primitive models by today's standards, when interindustry flows can be broken down into several hundred sectors.

Leontief's models were not ready for use in production planning during World War II, but as the expectations of victory rose—and worries about how the economy would react to reconversion to peace increased—two economists at the Bureau of Labor Statistics, W. Duane Evans and Marvin Hoffenberg, joined forces with Leontief in developing detailed input-output tables on the basis of the last Census of Manufactures, for 1939. In 1944 they triumphantly produced a 95-sector table, which BLS immediately put to work as the basis of a study of the production that would be needed from each industry if the economy were to achieve full employment by 1950. One thing the study clearly showed was that there would be a very high ratio of demand for steel in the production of consumer goods. The BLS figures indicated that the minimum amount of steel needed to ensure full employment in 1950 was 98 million ingot tons, higher than in wartime. There was some grumbling about this forecast, especially by heads of steel companies, who had resisted government pressures to expand capacity sufficiently at the start of the war and who were afraid of being caught with excess capacity after the war. But actual steel production in 1950 proved to be close to what the Leontief input-output model had predicted—96.8 million tons.

The input-output technique for production planning appeared so promising that the U.S. Air Force joined with the Bureau of Labor Statistics in what they called "Project Scoop," a new 200-sector input-output table based on 1947 data, at a cost of $1.5 million. Governments in many other countries began to construct input-output tables of their own. However, with the coming of the conservative Eisenhower administration in 1952, suspicion of input-output analysis arose on ideological grounds—the fear that such analysis was an entering wedge for socialist planning. Undersecretary of Defense Roger M. Kyes, a former General Motors executive, killed the Air Force's input-output project.

Not all the opposition to input-output has been ideologically motivated. Within the economics profession, Leontief has been criticized for not building sufficiently complex and dynamic models, ones that would be more sensitive to the effects of changing prices, interest rates, profits and money flows, technology, and other variables. The Leontief model is based essentially on fixed coefficients—that is, on fixed technological relations between what producers buy and what they sell. For instance, for forecasting purposes, a Leontief model would normally assume that (except for changes in steel production processes) steel producers would use about the same inputs of ore, coke, chemicals, and other materials in the period being forecast as in the year the latest input-output model was computed (which characteristically would be several years ago). But as prices change, consumer demand varies, or materials shortages develop, industry must shift the composition of its inputs and alter the composition of its outputs (e.g., the automobile industry's shift to smaller cars).

The assumption of fixed coefficients may be valid for small changes in output, but where output changes are large—for example, a major military buildup or cutback, or the introduction of a new industry into a developing country—fixed coefficients can lead to seriously faulty projections. Walter Isard and Phyllis Kaniss note that because each sector of the model is an aggregation of many individual firms, the coefficients derived for any one sector represent an average for the production operations of firms which may differ greatly in size, efficiency, and other factors; constant production coefficients may then lead to inaccuracies when changes in final demand affect firms differently, as is often the case.[7]

Static input-output models implicitly assume the existence

of unused capacity and resources, to allow for expansions of output when required. But if the economy is already operating at or close to capacity, a very different pattern of resource use may be required, since more capacity must be built if production is to be expanded.

Leontief has been deeply concerned about all these problems, and at least as aware as his critics of the need to dynamize input-output analysis and widen its practical applicability. In his more recent, more sophisticated models, for example, he has introduced capacity-building activities. He has constantly pressed for better data and for shortening the length of time it takes to draw up a set of tables as a base for forecasting or planning. He has studied technological change and, with the help of his staff, has sought to change coefficients whenever the data disclose the way industrial input-output relations are changing. Since the early 1950s he has been working on dynamic models that take account of stocks as well as flows of goods, inventories of goods in process as well as in finished form, capital equipment, buildings, and household stocks of consumer durable goods; all this obviously requires vastly more detailed and complicated data, and the use of linear differential equations where the static models use ordinary linear equations. He has also built interregional models to take account of flows of trade and investment among nations.

In the view of some, he has taken on a heroic but Sisyphusian task. The economic process, they say, is too complex and variable for such efforts at modeling. Leontief refuses to succumb to that sort of pessimism; and the very fact that input-output requires the consistent, orderly, and comprehensive collection of economic data has been a tremendous boon to quantitative economics.

Both the system's data needs and Leontief's tireless crusading for more and better information have led many nations to begin to coordinate their data-collecting efforts for their own purposes and for international comparative studies. Today, more than fifty countries have constructed input-output tables, including the United States, the Soviet Union, the Common Market countries, and Japan. Despite its imperfections and shortcomings, the input-output method, as a bold simplification of economic theory, is having a profound effect on economic analysis and planning.

V

LEONTIEF'S passion and goal is the rational use of economic planning for humane ends. He is not a collectivist or a totalitarian; he highly values individual freedom, the profit motive, and self-interest as steady and dependable spurs to business and individual initiative, and he prefers an open and democratic society. But he thinks it is increasingly essential to combine the profit motive with economic planning. As he puts it, "Under our system of free enterprise the profit motive is the wind that keeps the vessel moving. But to keep it on a chosen course we have to use a rudder."[8]

In the United States, planning has traditionally been considered good for families and businesses because, with limited resources, they must plan to achieve desired goals. However, it is generally believed that planning at the national level is bound to be inefficient and to interfere with the freedom of individuals and businessmen. Leontief insists that "the decision that confronts management is not that of how to choose between unrestricted competition and all-pervasive planning, but rather how to choose an effective combination of the two."

"Nothing," he has said, "is further from my mind than the notion that the profit motive, or, in a wider sense, self-interest, can be replaced by centralized decision making. . . . Milton Friedman repeats again and again that the profit motive is the powerful driving force that moves our economy, and I agree. But he also insists that we abandon the vessel to the whim of the wind, and let it go in the direction in which the winds happen to be blowing. Follow his advice and the ship will very quickly land on the rocks. I say: use the profit motive, but control it to move the economy in the direction you want to go. And to fix your course, you must have maps and charts."

He finds it ridiculous that the United States government should act in such an uncoordinated, short-sighted, and even self-contradictory way. The various regulatory agencies overlap but pursue the interests of their respective clients; the Council of Economic Advisers has only summary and approximate information about the economy and is a tool of the president that he can misuse or not use at all. Leontief thinks that if in the Council's place "we had a well-staffed, well-informed, and intelligently guided planning board, the mess in which the country finds itself today could have been avoided." The nation, he thinks, might have avoided the "hysterical" fuel shortage of 1973–1974, since it would not only have known about the shortage of refining capacity (which some few did know) but would also have been able to do something about it. Similarly, he feels there might have been a more sensible use of agricultural resources and less inflationary pressure.

"It is quite understandable," he says, "that business does not like any type of outside intervention. Business, however, is not as delicate a flower as its spokesmen would like us to believe." He thinks business would ultimately adjust to and benefit from the work of a planning board that would come up with valuable facts and projections, specific practical proposals, and long-range plans for government programs and policies which are now made carelessly or only implicitly.

The planning board's job, in his view, would be to outline

the choices, not set the priorities or choose the national goals. Congress would make the decisions. "Let the [advocates of competing interests] push for their particular solutions, but let us at least ascertain that each of these solutions can really be implemented in practice." For the short run, Leontief contends, the board should be looking three to five years ahead, but it should also be preparing for longer-run problems by peering ten to twenty years ahead, and beyond.

Perhaps as much as his critics—and doubtless more than some advocates of economic planning—he appreciates how incredibly difficult and complicated the planning operation would be. He recognizes that in some countries where it has been tried, planning has not worked very well. "Very often," he notes, "people say in some socialist countries it didn't work. Well, my answer is that many things these inexperienced, unimaginative foreigners cannot do, Americans should be able to do. We always pride ourselves on American know-how, enthusiasm, initiative. . . . We are a little like the Chinese. Somebody observed to me, in crossing from Hong Kong to Canton, look here, the Chinese run a very efficient capitalist system in Hong Kong and a very efficient socialist system on the mainland . . . I think this country has the ability to undertake new things and accomplish them. The joy of accomplishment should enable us to organize planning much more efficiently than the tradition-ridden Europeans could do it."

He thinks it would be quite useless to appoint a planning commission of, say, a dozen members. What is needed is not a committee but a planning organization—"and the board should have 200 analysts for each board member. It must be a large organization. . . . Here again, it is the custom now to denigrate bureaucracy. I find it very amusing to hear the president, who presides over a large bureaucracy in Washington, having all these unkind words to say about them. My feeling is that the American bureaucracy is a very good bureaucracy, very skilled, very dedicated. They really can solve problems. They have all the skills and moral qualities. If we have a terrific amount of graft, it was not among our bureaucrats. It was among the politicians. As a matter of fact, let me add that there is at least as much graft in private business as in politics. Private business is the main source of political graft, as we learn now. The most incorruptible part of our society, I would say, are our bureaucrats."

His planning bureaucracy would submit alternatives, just (he claims) as he does: "When I do my projections of the world economy, on the request of the United Nations, I don't make one projection. I make a whole spectrum of projections, so nobody can accuse me of being too conservative or too liberal or trying to spend too much on public health or too much on food. . . . I am very skeptical of the planner who says, 'Tell me what you want and I will deliver it.' Because it is like somebody coming to me and saying, 'I want to invite you to a restaurant; describe to me your tastes.' I say, 'Never mind, give me a menu, and I choose what I want.' I cannot even describe my tastes—the same applies to the public.

"Of course, there are certain aspects of a system toward which the final consumer is neutral. He doesn't want to choose between this type of motor and that type of motor; the structure of a motor does not affect him at all, only the result. But whether more people choose to work in agriculture or more in industry, whether people choose to live in the suburbs and travel long distances or live sacrificing some of the pace —this is not for the expert to decide, because this is a problem of cultural preferences, social preferences, individual preferences."

What about Leontief's own preferences? What type of economy would he prefer—say, fifty years from now? "It will be an economy," he answers, "in which the state plays a large part, but much will be left in private hands. If at present the government budget is 30 percent of GNP, it might be 50 percent or over. Certain branches, like public utilities, will be completely nationalized or strictly regulated."

The government will control the system at many points. "The Keynesian approach was essentially to choose a few instruments, two or three or four, and use them to bring about what you want. My feeling is that that is like trying to use a brake on one wheel. Steering always produces a strain, and the fewer points of control you have, the more pressure you put on certain parts of a system. Look at a skier who steers himself going in the slalom—he bends his whole body, he doesn't use just one ski to brake. If he does, he will just turn over."

Leontief has never been a Keynesian; he never felt that simply expanding and contracting aggregate demand could provide government with a means of coping with the many problems of an advanced industrial economy. However, he agrees that Keynesian fiscal policies have proved to be capable of providing "strong defenses against economic cave-ins of the type experienced in the early thirties." Yet the Keynesian approach is inflation-prone; it cannot achieve full employment without generating inflation. At minimum, he feels, the Keynesians will have to intervene in the price and wage process, and he approves of those (such as Galbraith) who recognize this: "The sophisticated, as contrasted to the stripped-down popular version of Keynesian theory, shows great concern for the crucial relationships between effective demand and fiscal policies, on the one hand, and prices and wage rates on the other. In particular, it demonstrates the importance of fixing either prices—at least some of the more important prices—or money wage rates, by custom, agreement or by edict."[9]

The most difficult part of the whole problem, in his view, is how to combine government controls with the pursuit of individual self-interest. He has little faith in voluntarism or altruism on the part of the citizenry. He believes President

Ford was "very naïve" to think he could induce Americans to use less gasoline by wearing a button to that effect. "In this sense I am in agreement with the conservative's view that it is self-interest which moves people, not belief in the common good. Sometimes people are inspired and ready to act in the common good and are willing to forget self-interest. But this lasts for only a short period of time. It is a little like when the physicist tries to produce a very high voltage—he can do it for only one quarter of a second, because if he tries to make it last too long, everything burns out. I think we can rely on people's self-sacrifice at critical moments for a short time—a couple of days or a couple of months. I just don't believe much in moral incentives. When I visited Cuba, I expressed my skepticism, and the Cubans were never mad at me. If I am not mistaken, their present system does not rely entirely on moral incentives."

The same thing, Leontief adds, is true in the United States. The profit motive in American enterprise is powerful, he says. "It induces American industry to use terrific amounts of know-how to accomplish results. It is so strong that I do not believe that if you put certain obstacles before it, to induce it to move in other directions, it will give up. I do not think regulations will reduce the spirit of private enterprise. Put a wall before the American entrepreneur, he will just go in another direction, I hope in the direction I want him to go."

He feels that the use of investment controls may be one of the most practical ways of trying to steer the American economy in the desired direction, and away from undesired directions. He also believes that "our financial system and even our political system is ready to accept certain kinds of monetary and credit actions." The business and banking community, he claims, are unwilling to rely entirely on "so-called free capital markets." This is particularly true for endangered or failing businesses or banks. But Leontief thinks that credit control without economic planning is blind.

VI

LOATH as he is to draw up utopian designs for society, Leontief's own values are expressed by the uses to which he has put his technical work. He very early realized that his input-output system could become an important tool in helping the less developed countries modernize their economies, and indeed input-output has become a useful component of development planning. He has written frequent papers on the subject. He has described the methods and procedures for using the tables to study the economies of underdeveloped nations and to project and evaluate development plans;[10] he has pointed to the need for even bigger capital transfers from the developed to the underdeveloped areas;[11] and he has warned that even with massive foreign aid or loans, a developing nation's economy will stagnate without intelligent direction of domestic resources.[12] As a realist, he understands that the very conditions which he is attempting to improve in the poor countries make it nearly impossible to collect data accurate and detailed enough for a precise analysis of their situations. For this purpose, he has proposed a Permanent International Scientific Institute for Research in Technical Economics, which would be similar to the International Atomic Energy Agency in Vienna.[13]

His longtime interest in the problems of pollution have coalesced in the UN study of the world environment. Besides his Nobel Prize speech, he has written several papers showing how input-output is a natural instrument for studying the costs of environmental pollution. He delivered a notable paper on the subject[14] at the International Symposium on Environmental Disruption, held in Tokyo in 1970. In that paper, he described how undesirable by-products and valuable but unpaid-for natural inputs can be incorporated into the conventional input-output picture of the economy by extending the coefficients to combine the amount of different pollutants generated with the output of each economic sector. The cost of depollution could be set up as a separate sector—in effect, a new industry—for the purpose of studying its costs and benefits. Says Leontief: "Once this has been done, conventional input-output computations can yield concrete replies to some of the fundamental factual questions that should be asked and answered before a practical solution can be found to problems raised by the undesirable environmental effects of modern technology and uncontrolled economic growth."

He has sought to use his all-purpose tool of input-output as a means of facilitating international disarmament. His first efforts, during World War II, were designed to show how disruptive or nondisruptive the decrease in military spending would be, and specifically where key problems would occur. His studies helped to counteract the predominant view among the Keynesian economists of the day that disarmament would tumble the economy back into depression. Increasingly antimilitarist after the Korean conflict and especially during the Vietnam war, Leontief focused his studies on measures to promote world peace in the 1960s. In 1965 he and four of his associates at the Harvard Economic Research Project gathered an immense collection of military data and published papers assessing the economic implications of hypothetical steps toward disarmament. Military spending, they concluded, is a useful crutch for a stumbling capitalist system; but Communist systems rely heavily on military spending for their own internal political reasons, as well as their external fears or designs. Leontief was looking for a rational way to ease the problems in both worlds and to mediate between them.

VII

LEONTIEF has a grand conception and grand goals for economics. "Locke, Hume, Adam Smith, and even Karl Marx and John Stuart Mill," he says, "were philosophers interested in economic problems among other things." Like them, Leontief is an economic philosopher. The particular lens through which he peers happens to be the discipline of economics, but it might have been any of the social sciences—or even journalism. His reports of trips to the Soviet Union, Japan, Cuba, and China display reportorial skill and insight of a high order; he has a sharp eye for the telling detail. From Tokyo he wrote:

> In contrast to the RCA television set in the New York hotel, the National set in the Tokyo room works perfectly. The soap and the deodorants advertised on the morning programs are virtually the same—however, the jerky actors of the animated cartoons have faces of Oriental cast. The panel on the side of the night table, full of buttons, controlling lights and other appliances, makes me feel like the pilot of a 707. . . . [Then he took a drive through the Tokyo dumps.] Mile after mile we drove through mountains of refuse, plateaus and valleys of refuse, traversed by caravans of garbage trucks and provisioned by fleets of garbage barges. The only living things in the dump were swarms of crows, replaced for the night shift . . . by hordes of rats. The color of this fantastic landscape changes from one stretch to the next, like that of a natural desert. Red, green, blue, and yellow plastic material, deposited in winding layers and fantastic strips, reminded me of a gigantic Jackson Pollock. . . . At one point, we came upon a weird muddy lake with a garbage bottom and garbage shores. At one edge, tilted on its side, was a small fishing schooner, the kind one sees at the fishing piers in Gloucester. This boat, however, was dead, its windows broken, gray paint peeling, planks missing in its decks.[15]

At the root of Leontief's philosophy is the belief that knowledge, empirical knowledge, to be gained with eyes and nose and hands and ears and measuring instruments of all types, should be the foundation of all theories and decisions. A failure of policy to attain its ends, he thinks, can usually be blamed on insufficient knowledge. Better knowledge might at least make it plain that certain ends cannot be achieved. But is such knowledge useful? He answers, "Yes. Even partial knowledge is useful, since it can protect us from at least some, if not all, mistakes."

He has been the scourge of many of his fellow economists for what he considers their elegant but useless theorizing (". . . a close and easy fit of the theoretical garment should count more than the elegance of its fashionable cut . . .")[16] and their penchant for prescribing measures which are not based on careful diagnosis of the actual or potential ills such measures are meant to cure. In his presidential address to the American Economic Association in Detroit in December 1970, he declared that the main fault of contemporary economics was not the irrelevance of the practical problems to which present-day economists addressed their efforts but rather "the palpable *inadequacy* of the scientific means with which they try to solve them."[17] He found it unfortunate that anyone capable of learning elementary or advanced calculus and algebra and acquiring an acquaintance with the specialized terminology of economics could set himself up as a theorist, while concealing the "ephemeral substantive content of the argument behind the formidable front of algebraic signs." The mathematical model-building industry, he said, has grown into one of the most prestigious branches of economics. Construction of a typical theoretical model can be handled now as a routine assembly job; but the assumptions on which the model is based are easily forgotten, and it is precisely the empirical validity of the assumptions on which the usefulness of the entire exercise depends.

He complained that preoccupation with this sort of hypothetical reality has led to a distortion of values in the academic world. Creators of complicated mathematical models are more quickly recognized for their genius than those who, through long and hard labor, create sturdier models based on observed facts. The scoring system that governs the system of rewards of the economics profession encourages the trend toward fuzzy thinking and scholasticism. But the weak and slow-growing empirical foundations of economics cannot support the proliferating superstructure of pure and speculative economic theory.

In his address, he lamented the lack of interdisciplinary cooperation in the social sciences. He found cooperation across the traditional frontiers which now separate economics from engineering and the social sciences hampered by his own field's comfortable sense of self-sufficiency, resulting from economists' "undue reliance on indirect statistical inference as the principal method of empirical research."

He wound up his Detroit speech with a plea to his fellow economists to try harder. The academic economists, he said, "were ready to expound, to anyone ready to lend an ear, our views on public policy. . . . We should be equally prepared to share with the wider public the hopes and disappointments which accompany the advance of our own often desperately difficult, but always exciting intellectual enterprise. This public has amply demonstrated its readiness to back the pursuit of knowledge. It will lend its generous support to our venture too, if we take the trouble to explain what it is all about."

More recently, he has become disheartened about his efforts to drag economists away from their arid theorizing. "Economics," he has said, "is like a big dead whale. We must get a tractor or a bulldozer and push. We will need a great many hooks, or it will come apart."

Has economics any future as a separate discipline? "In the long run," Leontief answers, "economics is slowly disintegrating. Operations Research will bite off a big piece of it—microanalysis. In the business schools they already are teaching the same things as the economics departments, but they are doing it right." He feels that economists are moving slowly back toward an appreciation of the importance of general equilibrium theory, which reached its low point early in the century under the influence of Alfred Marshall. But he still worries that economists will not work hard enough to get at the facts of society—"they will stand on their head to interpret stupid data." "Fortunately," he adds, "the world is stable enough to survive the ignorance of scientists."

Despite his tireless scoldings of his own profession, Leontief is by no means lacking in appreciation from his colleagues. "Oddly enough," writes Professor Robert Dorfman of Harvard,

> though he has contributed a major innovation, novelty and originality are not what Leontief seems to strive for in his work. . . . He made his discoveries by starting with a traditional concept that dissatisfied him and striving to bridge the gap between that concept and the observable phenomena that it purported to describe. . . . The ability to perceive flaws in concepts that have long been taken for granted is a precious and rare one in all sciences, and the "scientific" (or empirical) approaches exist side by side. Leontief has this ability to a superlative degree. . . . Thus, Leontief stands, near the end of his career, as the model of the scientific method in economics. I cannot think of anyone who excels him in this regard among living economists.[18]

VIII

IN 1975, just a year or so from retirement, Leontief left Harvard with a bang. He departed for New York University in a rage at Harvard's own economics department, complaining that it had never given enough moral or intellectual support to his work, had not appointed or given tenure to his disciples, and had refused to broaden its scope by granting tenure to any radical economists.

Actually, Leontief's politics are rather difficult to define. He has become more liberal—or radical—as he has matured. However, his political ideas seem to fluctuate over time. He supported and consulted with Senator George McGovern in his 1972 presidential campaign; however, he played no part in McGovern's ill-fated plan for income redistribution. He held that the need for a fairer distribution of income would become the dominant political issue of the last quarter of the twentieth century. But more recently he has focused his attention on the need for economic planning rather than income redistribution, contending that it would be too difficult and perhaps politically dangerous to use planning simultaneously to make the economic system function better and to redistribute income in the short run.

With Leonard Woodcock, president of the United Auto Workers, Leontief in 1975 became cochairman of the Initiative Committee for National Planning to promote the cause of establishing a central government planning board in the United States. The initiative attracted widespread national interest, from some liberal businessmen as well as politicians. But Leontief has never been a big "growth man," or a champion of growth for growth's sake. At the time of the Humphrey-Nixon presidential campaign in 1968, he wrote: "The average American, engulfed by the rising flood of bigger and better cars and taller high-rising buildings unloaded on him by private enterprise, has about as much chance to opt against the wholesale destruction of the amenities of private civilized life, as he will have to express his preference for the cessation of our military intervention in Vietnam by choosing freely between Hubert Humphrey and Richard Nixon."[19] While the subsequent years have not done much to dispel his worries about the inability of the average Ameri-

can to turn the industrial process toward better social advantage, he clings to hopes that planners could steer economic
growth toward healthier and more satisfying ends than self-
interested corporate leaders have in the past.

He believes that most of our growth in recent years has
been brought about by technological development rather
than increased capital investment, and he would provide
maximum support to the development of new knowledge. He
has argued that the product of industrial research, which is
new scientific knowledge or technical know-how, differs from
most other goods in one respect: "It can be useful, it might
turn out to be useless, but it cannot be used up." As he wrote
in 1960,

> not only can the same person make use of an idea, of some
> specific piece of technical information, over and over again
> without the slightest danger of exhausting it through wear, but
> the same idea can serve many users simultaneously, and as the
> number of customers increases, no one need be getting less of
> it because the others are getting more.[20]

He concluded that whatever the method of financing the costs
of gaining new knowledge and of rewarding the producers of
that knowledge, the economic benefits of scientific and industrial research can be exploited fully only if no one—no one at
all—is prevented from using its results by the price which he
has to pay to do so. Knowledge thus becomes the premier
public good.

Education is an essential part of the process of expanding
and transmitting knowledge:

> Education is a consumer good satisfying one of the most im-
> portant human needs; at the same time it is a productive social
> investment leading to an increase in future material output.
> Thus it raises the standard of living of the present generation
> while contributing indirectly to the income of future genera-
> tions as well.

The thought that greater economic equality might be justifiably sought not only among the members of the same society, but also among different generations, admittedly runs counter to the deep-seated drive for growth and progress. But isn't this just another instance of a continuing struggle between atavistic, intuitive drives, on the one hand, and enlightened rational value judgments, on the other?[21]

There are ambiguities and contradictions in Leontief's thought—notably in his struggle to find ways of reconciling equality and progress, freedom and intelligent social control and direction. Had his own life's work not been so heavily devoted to technical work on input-output, he might have had a better chance of establishing a "school" of disciples. He says, "Prophets should be ambiguous. Like in the Bible. You find a lot in common between Keynes and Marx—both were sufficiently ambiguous to build schools. If someone is absolutely clear, he cannot make a school."

Despite his sympathy for socialism and his commitment to planning, Leontief is not a Marxist. Indeed, he points out that Marx sometimes wondered whether he was a Marxist himself. (Similarly, some of Keynes's latter-day admirers have wondered whether Keynes was a Keynesian. Even President Richard Nixon, in what may go down as his most famous economic statement, said "I am now a Keynesian." As for Leontief, he has always considered himself anti-Keynesian.)

He admires Marx as a source of what he considers direct observation of the capitalist system, and he recommends the three volumes of *Capital* as excellent source material on capitalism. From those three volumes, he says, you get more "realistic and relevant first-hand information than you could possibly hope to find in ten successive issues of the United States Census, a dozen textbooks on contemporary economic institutions and even, may I dare say, the collected essays of Thorstein Veblen."[22] He thinks Marx had a "great feeling for social structure and the social order—what the capitalist system is."

But he thinks little of Marx as an economic theorist. "Marx's theory," he says, "is classical economics, slightly socialized." Marxism, argues Leontief, is not a theory of a centrally guided economy but a theory of rampant private enterprise; whatever references Marx made to the economy of the coming socialist order were "brief, quite general, and extremely vague."[23]

Leontief holds that the modern theory of prices owes nothing to the Marxian version. However, he thinks *Capital* helped more than any other single work to bring the whole problem of business-cycle analysis to the fore in economics; he cites evidence to show that toward the end of his life Marx actually anticipated the mathematical approach to business-cycle analysis.

He feels that Marx's analysis of the long-run tendencies of the capitalist system led to some impressive prophecies: the increasing elimination of small and medium-sized enterprises, the progressive limitations on competition, the incessant technological progress and ever-growing importance of fixed capital, and the undiminishing amplitude of business cycles. But he finds Marx wrong in his prophecy of increasing impoverishment of the working class.

Marx, he finds, was a great intuitive "character reader" of the capitalist system. His rational theories do not hold water; their inherent weaknesses show up as soon as other economists, not endowed with Marx's exceptional sense of reality, try to proceed on the basis of his blueprints. Leontief compares Marx to an experienced layman with an insight into character—a kind of witch doctor—who can predict individual behavior more accurately than can a professional psychologist. Most importantly, Marx illuminated the role of conflicting economic interests as driving forces in the social and political process. But, says Leontief, "His followers . . . never made good on the vast claims staked out in that brilliant flash of insight along these inaccessible frontiers of knowledge: neither, one hastens to add, has anyone else."[24]

Leontief has sought to bring his own fresh eyes to bear on the reality of socialism and Communism as it has been developing, much as Marx did with capitalism. Though sympathetically interested in socialist planning, he has been unsparing in his criticism of Soviet practice: "So far as the Russian technique of economic planning is concerned, one can apply to it in paraphrase what was said about a talking horse: the remarkable thing about it is not what it says, but that it speaks at all."[25] Soviet economics has remained static and sterile, "a huge, impassive, immovable monument to Marx— with scores of caretakers engaged in its upkeep, fresh flowers placed in slightly different arrangements at its feet from time to time, and lines of dutiful visitors guided past in never-ending streams."[26]

In Cuba in 1969, Leontief decided that the Castro government had made a wise decision in switching its emphasis in planning from industrial back to agricultural development. Land reforms had made the landholders uninterested in the back-breaking job of cutting sugar sixteen hours a day. So the cane growers had to fall back on inefficient and unskilled volunteers from the city; this had led to a lag in sugar production. The country was having the same problem in harvesting other crops, with labor being drained off the farms by the increased size of the armed forces, more years spent in school, and the enhanced attractiveness of the cities, where wages were higher. Yet he left Cuba with an upbeat impression: the performance of its socialized economy was not inferior and possibly even somewhat superior to that of the Communist states of Eastern Europe in the early stages of their development. Of course, some individual operations were "shockingly inefficient," the bureaucracy was "often very clumsy," and continued reliance on "moral" instead of material incentives was adversely affecting the productivity of labor. Subject to those constraints, however, Leontief found the overall direction of Cuban economic policies "intelligent and aggressively imaginative." He concluded his August 1969 report with a thrust at United States foreign policy toward Cuba: "The policies of non-recognition and economic boycott isolate not Cuba but the United States."[27]

By 1971, however, his enthusiasm had flagged somewhat. What was to have been a 10-million-ton harvest had fallen short. In a critique of Castro's speech on the failure of the harvest, Leontief said it was not—as Castro contended—bad planning that had caused the debacle but the "characteristically low production of labor, rooted in the basic difference between a socialistic and an individualistic society."[28] The moral incentives for hard work had proved to be less effective than the threat of joblessness and hunger or the rewards of higher wages, salaries, and profits. Careless handling of harvesting machines had undermined any help they might have given as substitutes for labor. All the evidence indicated to Leontief that moral incentives had again failed to measure up to more conventional, individualistic self-interest. Yet material self-interest had persisted, in this case weakening and distorting the hoped-for effects of moral motivation.[29]

Three years after his excursion to Cuba, Leontief finally made it back to Mainland China, the scene of his first adventures as a budding economist and planner. After observing the animated life of five Chinese cities, visiting factories, schools, and communal farms, watching the Chinese countryside glide by on a long railroad journey through the eastern provinces, Leontief announced (in the famous words of the newspaperman Lincoln Steffens), "It works."[30] He was much impressed with the cleanliness everywhere, the lack of hungry, sick-looking people, the efficiency of the railroads, and the lack of waste (even the grass in the parks was cut for hay).

From his own observations and from discussions with members of the Economic Institute of the Academy of Sciences in Peking, Leontief concluded that the management of the economy, as reconstituted after the Cultural Revolution, "is circumspect and careful rather than bold and experimental." Chinese priorities were the reverse of those in other socialist societies and in many other developing countries: in China the emphasis was on agriculture first, then on light industry, and lastly on heavy industry. Even so, he said, the growth rate was estimated to be as high as 7 percent a year. All prices were held absolutely constant, since Chinese planners did not want to disturb consumers by erratic changes. Rice, grain, cooking oil, and cotton cloth were rationed, but the workers were able to obtain all the basic human needs— which was all the planners had promised to deliver. The Chinese have been successful in evading the famines of the past—they plan for two good crops, two bad ones, and one disastrous one every five years! Their population control efforts are working in the cities, says Leontief, but are having little effect in rural areas.

He finds that China has the usual problems that come with collective action: incompetence, inefficiency, and misrepresentation by the leaders, and a tendency among the organizers to underestimate the total effort required to do a job. Although the workers seem to be cooperating on the whole— with the help of iron discipline and threats of severe punishment—there is still a tendency for many to soldier on the job, and "the steady flow of refugees over the Hong Kong border testifies to the fact that the rigors of communal discipline prove to be unbearable to many of the people."[31] Leontief prophesies that as the system grows more complex, it will become increasingly hard to manage. And as the basic needs of the people are more fully satisfied, they will want more freedom of choice, particularly in the realm of ideas. "In the long run," he concludes, "freedom, or to put it more positively, the open conflict of ideas would prove to be an indispensable condition for human progress and social advance.

Not unlike food, freedom is a source of direct personal satisfaction, but it is also an indispensable condition of health and normal growth."[32]

I X

THUS, Leontief is a man without illusions, a planner who will not abandon his love of human freedom, a socialist who knows the normal limits of altruism and the powerful force of ego and self-interest. Leontief sees the Soviet experience as an object lesson in how socialism can turn sour.

In the Soviet Union, he feels, the Communists believed that history had given them a mandate to transform society and build an industrial economy. Because that transformation demanded a powerful administration to plan and control it, a large bureaucracy was required, with an inner group to run the bureaucracy. That bureaucracy was transformed, as Milovan Djilas puts it, into a "new ruling class," and socialism gave way to dictatorship. The new class did not possess the traditional power bases of the bourgeoisie or the aristocracy; legally it owned no land or capital, since these "belonged" to the people as a whole. Thus, it needed to develop an ideology to make its rule legitimate. As with all groups that rise to power, the bureaucracy's first goal was to maintain itself and to increase its power and security. This resulted, Leontief believes, in an "ideological economy" in which economic progress and individual freedom were subordinated to the political needs and theories of the ruling class, which claimed that it knew how to engineer production efficiently but actually knew only how to seize and maintain control of the economy.

Communist policies intended to ensure rapid industrial growth were turned to serve the interests of the party and its leaders. The original ideal of socializing the economy was subverted; under Stalin the Soviet Union became a tyranny of monopolistic control. Workers became prisoners of a closed system. Strikes were outlawed. The emphasis on heavy industry and military expenditures enhanced the stability and prestige of the regime; the secret police ensured its dominance. Rigid control of the working class had become a need of the new class, perhaps its guiding principle.

Under Stalin, economic science in the Soviet Union declined, says Leontief. The ten years before the first five-year plan had been marked by lively economic discussion, but the true economists vanished without a trace, victims of the secret police. The Communist leaders became their own economists, only to follow the precepts of capitalism's Founding Fathers, Benjamin Franklin and Adam Smith: "To expand one's income fast, one must channel as large a part as possible—and then more—into productive capital investment. This means consumption must be restricted." In other words, keep the workers working hard while holding down their living standards—a system Marx described in pejorative terms. Soviet planning was improvised and clumsy, and there were abundant examples of miscalculation: too much ore mined and not enough coke made to produce the planned amount of steel, lack of spare parts, wrong choices between gas and coal for energy supplies, etc.

But Leontief finds that since Stalin's death, the Soviets have been struggling to abandon their rigid ideological approach to economics. They began to publish articles (mostly critical) about American economic theory in the early 1950s; and by the end of the decade the criticism had lessened. They began to talk about input-output, and even claimed to have invented it. When Leontief visited the Soviet Union and lectured at the University of Moscow, they presented him with a handsome leather-bound copy of the Russian translation of his first paper. Leontief found the Russians well acquainted with the fundamentals of his system, learned that they had begun requiring mathematics in the economics departments of their universities, and had established chairs in econometrics at the universities of Moscow and Leningrad. They were in the process of building a new center for advanced economic research, with large-scale computational facilities, at Novosibirsk in Siberia. His hopes stirred: perhaps ways could be found to combine socialist values and individual freedom through the mediation of economic science. He has still not abandoned that hope, and almost certainly never will.

Kenneth E. Boulding

The Economics of Peace and Love

"I, too, dislike it: there are things that are important beyond all this fiddle. Reading it, however, with a perfect contempt for it, one discovers in it, after all, a place for the genuine."
—*Marianne Moore*

I

KENNETH BOULDING has never been content to live within economics as he found it, a rather narrow, obsessively rational discipline, in which people are treated as gray abstractions and from which God is excluded. All his life, Boulding has striven for a more adequate means of understanding human behavior and the ethical content of human action.

Any serious effort to explain human behavior can neither exclude particular ways of knowing nor, for the sake of doctrinal purity or coherence, too finely distinguish among them. Human action, as Boulding sees it, can be comprehended only through the entire range of our knowledge—rational, instinctual, and mystical.

As a young man, Boulding was deeply troubled by the dichotomy he felt between spiritual and worldly reality:

> *Not what we think, but what we do has been*
> *The standard of the world: so have I tried*
> *To wall out God with deeds. And yet inside*
> *My soul blazes His light despite my screen.*[1]

Boulding's life can be seen as an attempt to bring the two sides of his identity into harmony. In his work he has sought to move beyond economics toward a general social science that would deal with man in his totality. But at the same time, he has increasingly emphasized the practical uses of religion. He has seen the object of meditation as not so much to achieve union with the divine as to receive instruction from the divine, "and very practical instruction at that." He has

[191]

felt an indissoluble bond between the task of understanding human behavior and the task of improving it.

To his fellow Quakers and other religious believers, he has stressed that goodwill is not enough; concrete knowledge of how society functions is necessary if high moral objectives are to be achieved. And to his fellow social scientists he has pointed out the need for religious contributions to social analysis. He does not believe that economists can rest neutrally with the question "How do people get what they want?" He contends that it is not even ethically neutral to help people get what they want. Rather, as social science develops, Boulding sees the critique of ends as becoming more and more important. The question, "Do I (or does anybody) want the right things?" becomes inescapable. But it is precisely the critique of ends, says Boulding, "which is the great moral task of religion."

Not only, in Boulding's view, is religion necessary for social science, it is also necessary for the social scientist himself. Because religion is part of the whole experience of life, the social scientist who does not participate in it "is cut off from a deep and meaningful area of human experience and is in this sense maimed."

II

KENNETH BOULDING might well have become a preacher instead of an economist; in fact, he is both. He was born in Liverpool, England, on January 18, 1910, the only child of William Boulding, a gas fitter, and Elizabeth Rowe, the daughter of a blacksmith who was also a lay preacher in the Wesleyan Methodist Church. Both of Boulding's parents were deeply religious. They had met at a Wesleyan chapel in London, and when they moved to Liverpool the nonconformist church remained a central part of their lives. His father also became a lay preacher and served for many years as superintendent of the local Wesleyan Sunday School. William Boulding was, however, a poor businessman, and from the beginning of World War I the family became increasingly hard-pressed financially.

Kenneth was a precocious child, but his precocity was at first hidden under a severe stutter. Though able to read at the age of three, he was considered retarded because of his speech defect by the headmaster of the first school he attended. When he was nine years old, he transferred from the Anglican St. Simons to the Unitarian Hope Street School. By then his stutter—which he was never entirely to lose—had improved, and his talents were quickly recognized. Three years later he won a scholarship to the prominent Liverpool Collegiate School.

While still in primary school, Boulding had begun keeping diaries and writing poems; in this he was encouraged by his mother, who had literary interests of her own. He read voraciously. *Alice in Wonderland* and *The Swiss Family Robinson* were early favorites of his. He later became absorbed in the works of George Bernard Shaw and H. G. Wells.

At first he followed his parents' religion. During his adolescence he spent several summers at a Methodist summer camp, which he found both spiritually and intellectually inspiring. This led him formally to enroll as a member of the Brunswick Methodist Church in 1926. But he then began to sense a lack of immediacy in the religion of his fellow Methodists.

Reading an account of conscientious objection during World War I, Boulding became interested in Quakerism. He sought out members of the local Quaker community and, without leaving the Methodist church, started to attend the Liverpool Friends Meeting. Initially drawn to the Quaker religion by his deep abhorrence of war, he soon found equal appeal in its form of worship and system of belief. The silent meetings, unencumbered by preachers' exhortations and admonitions, gave the young Boulding both the unmediated contact with the divine and the profound sense of community with other worshipers that he desired. As the Quaker thinker Howard Brinton has written,

> the mysticism of the Quakers is directed both toward God and toward the group. The vertical relation to God and the horizontal relation to man are like two co-ordinates used to plot a curve; without both the position of the curve could not be determined.[2]

Boulding also found support for his underlying optimism in the Quaker denial of predestination and its assertion of the perfectibility of man.

At Liverpool Collegiate, Boulding concentrated in mathematics and the natural sciences, excelling in chemistry. Having failed in 1927 to gain a scholarship to Cambridge, he won an Open Major Scholarship in the Natural Sciences at New College, Oxford, in 1928. But Oxford was not the Eden which Boulding had imagined. He felt excluded from the life of his college because of his social and educational background; he was hurt and angered by class discrimination. He felt less in common with the Quakers at Oxford, who tended to come from upper-middle-class homes and private boarding schools, than with the Methodists, who were generally scholarship students like himself. With these Methodists he joined in missions of evangelism and social service to working-class communities in the area.

Boulding's poetic muse also suffered at Oxford. He failed to win the much-coveted Newdigate Prize for poetry, which he had been determined to secure. And finally, in his first year, Boulding grew increasingly unhappy with the study of chemistry. By the end of the spring he had decided to switch to the study of the "modern greats" in the honor school of politics, philosophy, and economics. Despite the change, New College allowed him to keep his chemistry scholarship.

Thus, in June 1929, at the age of nineteen and a half, Kenneth Boulding confronted economics for the first time. He went to Lionel Robbins, then about to leave New College for a professorship at the London School of Economics, and asked for a reading list. Boulding later recalled their meeting:

> Lionel Robbins very probably has no recollection of this incident, but I have a very clear recollection of him sitting in the window seat in his room in the garden quad, illuminated by the watery Oxford sun, and this shy, gauche, undergraduate from Liverpool asking him what he should read in the summer in economics. I had never even heard of economics before and had not the slightest idea what it was all about. I got out a pencil and paper and Robbins drawled cheerfully, "Well, you might read Marshall, *Principles of Economics*, Pigou, *The Economics of Welfare*, Cassell, *Theory of Social Economy*, and Hawtrey, *The Economic Problem*." I wrote these books down, never having heard of them, went to the library and got them out. They seemed rather large, but I had a whole summer to read them in, so I went back to Liverpool and read them. I came back to Oxford the next October to find Henry Phelps-Brown installed as the economics tutor. He promptly proceeded to give me a little examination, which in those days was called a collection, in which I scored an alpha, the Oxford for an A. Obviously economics was something that I could do, and I have continued to do it on and off ever since.[3]

His spirits renewed, Boulding waded happily into the new field. His efforts were well rewarded. In 1930 he won the Webb-Medley Junior Scholarship in Economics, and the next year he was one of ten students to graduate with first-class honors in modern greats; he gained the best "first" in economics. A succinct paper he had written on displacement costs was accepted by John Maynard Keynes for publication in the *Economic Journal*. This very first paper, brief as it was, illustrates the lucidity and freshness of Boulding's mind. The concept of displacement cost—the notion that the cost of acquiring or producing any good was equal to the cost of *not* using the same resources to acquire or produce some other good—was sliding too easily and carelessly into mathematical economics, Boulding pointed out. The concept held with exactitude only in particular "displacement systems" in which the total quantity of resources were fixed and homogeneous, and in which definite quantities of only two products were produced by definite quantities of resources. But these were not the conditions of the real world. Productive resources— the "philosopher's stone of muddled economists," said Boulding—were not homogeneous:

> We cannot add a unit of labor, a unit of capital (whatever that may be), and a unit of land, with perhaps a little entrepreneurial ability thrown in as seasoning, and expect a fine pudding composed of homogeneous "units of productive resources."[4]

In any case, the simple logic of displacement costs would hold only for a point in time; it would break down over a longer period of time, since resources are really not fixed in quantity. Hence, the analysis of economic processes could not be built on the crude model of a fixed cake, in which "the bigger the slice which Johnny gets the less there is for Susie and Jimmie. . . ."[5]

Kenneth Boulding had found his profession.

III

IN POLITICS, he began as a socialist. While still in secondary school, under the influence of Shaw and Wells, he was drawn to the socialist platform of the Labor Party. At Oxford he joined the Labor Club and contributed to an undergraduate socialist journal called *The Plan*. Jubilant at the victory of the Labor Party in the election of 1929, he was appalled by its defeat in 1931:

> This disastrous election! The work of thirty years undone in a day by the stampede of unintelligent voters. It makes me weep. . . . The day of Social Reform is past: we have come up against the Rock of Private Property, and we or it busts, to put it concisely if ungrammatically.[6]

Yet by 1931, Boulding was drifting away from socialism, largely as a result of reading Karl Marx. The graduate student, confident of the clarity of his own reasoning powers and knowledgeable about the current state of economics, found himself forced to reject that cornerstone of Marxian economics, the theory of surplus value, which was based on David Ricardo's long discredited labor theory of value. This theory postulates that the value of a commodity is proportional to the amount of labor going into its production. Boulding saw this proposition as simply bad economics, incapable of sustaining a revolutionary doctrine.

Even more upsetting to Boulding was Marx's vision of historical change; the harsh dialectic of class conflict and brutal revolution repelled the young pacifist. At a time when many young intellectuals in England were starting to look with sympathy upon the Communist regime in the Soviet Union, Boulding was edging out of the socialist camp. In an article for *The Plan*, he urged that the transition to socialism be made very gradually—only after a long period of schooling the masses in their responsibilities under the new order.

After receiving his baccalaureate degree, Boulding obtained a renewal of his economics scholarship and spent an additional year at Oxford doing postgraduate work. Then, in 1932, he won a lucrative Commonwealth Fellowship for two years of study abroad. That fall he traveled for the first time to the United States and took up residence at the University of Chicago. At Chicago, Boulding studied with Frank Knight, the dean of American conservative economists. He found Knight stimulating and learned a great deal from him, particularly regarding the theory of the firm, Knight's specialty. But Boulding disliked the fragmentation of the American university curriculum into innumerable short courses and examinations. From the beginning, he was struggling to see and develop the unity of knowledge, and he resented its being parceled out and marketed on separate counters.

But Boulding was fascinated by the new country. In the summer of 1933, he set out with a couple of friends on an automobile tour of the American West. The trip was cut short when he received word that his father had died. Back in England he disposed of the bankrupt family estate and settled his mother with a relative. Then, depressed, he sailed off for the second and final year of his fellowship in the United States. He spent the fall semester of 1933 at Harvard studying with Joseph Schumpeter. From the Austrian émigré who had once expressed the desire to be "the best horseman in Vienna, the best lover in Europe, and the best economist in the world," Boulding acquired a tolerance for monopoly, an appreciation of the importance of entrepreneurial skill, and a conviction that the great problem of capitalist society was its inability to elicit people's moral allegiance. The semester at Harvard was suddenly interrupted in early December when Boulding suffered a collapsed lung. His mother arrived around Christmas to look after him and together they went out to Chicago for the remainder of the academic year.

In the summer of 1934, his fellowship expired, Boulding was back in Liverpool, faced with the necessity of finding a job. Britain was in the depths of depression; there were few academic positions available. He finally obtained an ill-paying post as an assistant at the University of Edinburgh. There he spent the next three years, living with his mother in a small flat overlooking the Firth of Forth.

For Boulding, the department of economics at Edinburgh

was both institutionally and intellectually unpalatable. A rigid academic hierarchy severely limited communication between junior and senior faculty members; the atmosphere was formal and, to those at the bottom, oppressive. Even worse, the economics being purveyed there was outdated. The mathematical techniques which Boulding had learned at Chicago were frowned upon.

This galled Boulding, especially since he had recently become accustomed to discussing economic questions with leading economists as an equal. At Chicago, Knight had encouraged debate, and Boulding had responded enthusiastically, taking up the cudgels on behalf of the concept of the *period of production*. One of the earliest notions about capital was that it constituted a fund for supporting other factors of production over a period of time. Boulding had sought to modernize the concept of the productive period, which he thought had relevance to an understanding of the changing economic process. To his delight, the debate on the issue he had raised was now being carried on in the economic journals. As far as Knight was concerned, the whole notion was useless; capital was too homogeneous and economic events too complex for the discussion of discrete periods of production to be meaningful. His response to Boulding's article, "The Application of the Pure Theory of Population Change to the Theory of Capital," was an article entitled, "Mr. Boulding and the Austrians"—heady stuff for a young economist. Boulding in turn issued a reply, and then published a couple of other articles incorporating a time variable into the theory of investment. While sticking firmly to the belief that time intervals were important considerations in production and investment processes, he finally conceded that, in all probability, "any physical 'average period of production' is not capable of measurement." In due course the debate subsided, and Boulding moved on to other economic questions. The experience, however, had been exhilarating, and he resented his treatment at Edinburgh as an immature underling.

But if his academic life was unrewarding, Boulding was finding fulfillment in religious activities at Edinburgh. Since his Commonwealth Fellowship, he had again drifted away from the Methodist fold; in Chicago he had been an active member of the 57th Street Quaker Meeting. Upon moving to Scotland, he became deeply involved in the life of the Quaker community there. He organized a Young Friends group and participated in various pacifist organizations, including the Friends Peace Board and the Peace Pledge Union. He also helped to set up a Quaker work camp, recruiting young people for manual labor and discussion sessions in impoverished Scottish communities. Finally, in 1937, he was sent as a representative of the General Meeting of Scotland to the Friends World Conference in Philadelphia. This proved a turning point in his career; while in America, he was offered an instructorship at Colgate University. He accepted, and made plans to emigrate.

Boulding's decision to leave Great Britain was hardly surprising. His assistantship at Edinburgh had expired, and there was little chance of its being renewed; a speech of his, attacking the Scottish university system, had aggrieved his superiors. Attempts to gain fellowships at Oxford had failed, and the job market was still extremely tight. All in all, he was happy to seek his fortune in the United States.

Boulding was delighted with his new surroundings. His colleagues were congenial, and the pleasant village of Hamilton in upstate New York was a happy change from the cramped urban quarters of Liverpool and Edinburgh. In 1938 he acquired a little house and sent for his mother. Both Bouldings were warmly welcomed and immediately accepted into the life of the small community. Mother and son felt that the American dream had, for them, come true.

Once at Colgate, Boulding embarked on the great task of writing an economics textbook. His reason for wanting to engage in such a project is not hard to find; it went beyond the simple desire to make money. More than other fields,

economics has revolved around a highly distinguished series
of textbooks. The first and greatest of these was Adam
Smith's *Wealth of Nations*; appearing in 1776, it held sway
for close to fifty years. Subsequently, the great debate between David Ricardo and Thomas Malthus was resolved by
the victory of Ricardo's *Principles of Political Economy and
Taxation* over Malthus's *Principles of Political Economy*.
The next generation of economists in the English-speaking
world grew up on John Stuart Mill's own *Principles of Political Economy*, which appeared in 1848. In the early twentieth
century, economists—including Boulding himself—were
weaned on the *Principles of Economics*, by Alfred Marshall.
All of these impressive tomes sought to demonstrate the essential coherence of the whole range of economic life by applying to it a relatively small but increasingly sophisticated set
of analytical concepts and techniques. The prospect of painting the total picture anew has understandably appealed to
many economists, including the young Kenneth Boulding
and, shortly afterward, to Paul Samuelson. Samuelson's text
carried off the major prizes, and swept the world as the new
economics. Yet Boulding's text was also a worthy achievement in rigorous analysis and clear presentation—as far as it
went.

Up to the point of writing his text, Boulding had been a
nearly pure theoretician. In a decade of studying economics,
he had done but one piece of empirical or public policy research: a study of the problems of British meat and milk
production. In a somewhat pompous preface, he made a virtue of his predilections:

> It is my belief that a work on principles should compete with
> neither the popular magazines nor the encyclopedias. Conse-
> quently I have not endeavored to write a compendium to cur-
> rent economic problems, for the reason that by the time the
> student has to face economic problems those of today may no
> longer be current. . . . It also seems to me that it is unwise to
> crowd a principles course with masses of factual material in

special studies—labor, marketing, etc.—merely for the sake of giving the work an air of factuality. The place for such factual studies is later in the student's career, when he has acquired the techniques for interpreting the monstrous riddle of the factual material.[7]

Economic Analysis was thus organized not according to subject matter but according to the mode of analysis employed. The first half of the book relied on the concepts of supply and demand as the principal tools for analyzing price determination and distribution, as well as the theory of money, banking, international trade, and the business cycle. The second half brought in the more sophisticated marginal analysis, showing how it underlay the supply and demand curves and using it to explain the theory of the firm, consumption, imperfect competition and monopoly, and the formation of capital. Although he did not push his case too strongly, Boulding continued to stress the importance of incorporating a time variable into the theories of production and of the firm; his concerns were developed in two chapters, "Time, Production, and Valuation" and "The Equilibrium of an Enterprise in Time." Boulding campaigned in his text for the American economist Irving Fisher's "equation of exchange"; he chided his fellow economists for their unwillingness to recognize the usefulness of Fisher's concise expression of the relationship between the supply of money and its velocity of circulation on one side, and the price level and volume of economic transactions on the other.

There was one great curiosity about Boulding's text: like Sherlock Holmes's hound of the Baskervilles, it did not bark. There is no mention of Keynesian macroeconomics in Boulding's first edition of *Economic Analysis*, though the book was published in 1941 and Keynes's *General Theory of Employment, Interest, and Money* was published in 1936. This omission merits explanation. Boulding, in fact, had every reason for not immediately jumping on the Keynesian bandwagon.

As an undergraduate, he had been at Oxford rather than at Keynes's Cambridge. In the United States, Boulding had studied with Frank Knight, who was to become a leader of the opposition to Keynes; indeed, the whole "Chicago School" fiercely resisted the Keynesian revolution and, when the revolution had apparently succeeded, sought to stage a counter-revolution against it. At Harvard, the center of the Keynesian revolution in America, Boulding had studied with Schumpeter, who also refused to serve under the Keynesian banner. At Edinburgh, an academic byway, Boulding had had little exposure to the intense debate that Keynes had provoked elsewhere. But, ultimately, Boulding takes responsibility upon himself for simply being unable to make sense of the book when it first came out. Many years later, recalling his problems with Keynes's difficult *General Theory*, Boulding said, "It's an appalling book. Keynes didn't understand everything he was talking about. It shows an immensely creative muddle-headedness. I certainly didn't understand it." Boulding therefore remained aloof from the great debates, proselytizings, and sortings-out of the late 1930s, writing his book as if the great revolution in modern economic thought were not going on outside his window. The dust cleared in time for the second edition; when a renovated *Economic Analysis* came out in 1948, it was a Keynesian textbook.

I V

WHILE Kenneth Boulding the economist was working on the first edition of his essentially conventional textbook, Kenneth Boulding the British expatriate, poet, and devout Quaker had been going through the most serious spiritual crisis of his life.

The rise of Fascism, culminating in the outbreak of war in

Europe, had severely shaken the pacifist movement in Great Britain and the United States. Many pacifists abandoned their opposition to war in the face of Nazism, Hitler's persecution of the Jews, and his savage assault on other nations and on the basic values of Western civilization and religion. How was it possible, in the name of humanity, to remain aloof from the greatest struggle against inhumanity ever waged?

Boulding found his own pacifism undermined. Increasingly, he found himself unable to love his enemies. He felt overwhelmed with anger:

> *I feel hate rising in my throat.*
> *Nay on a flood of hate I float,*
> *My mooring lost, my anchor gone,*
> *I cannot steer by star or sun . . .*

> *I hate! I hate! I hate! I hate!*
> *I hate this thrice-accursed State,*
> *I'll smash each blooshot German face*
> *That travesties the human race!*[8]

On May 15, 1940, a mystical experience reestablished Boulding's faith in the pacifist ideal. Emerging from his bath, he saw Christ suffering on the cross, assuming the sins of all mankind, of people no better than the Germans, no better than Boulding himself. His hatred disappeared. Once again, Boulding felt kinship with all men. His faith in nonviolence was renewed.

> *Hatred and sorrow murder me.*
> *But out of blackness, bright I see*
> *Our blessed Lord upon his cross.*
> *His mouth moves wanly, wry with loss*
> *Of blood and being, pity-drained.*
> *Between the thieves alone he reigned:*
> *(Was this one I, and that one you?)*
> *"If I forgive, will ye not too?"*

Kenneth E. Boulding

My vial of wrath breaks suddenly,
And fear and hate drain from me dry.
There is a glory in this place:
My Lord! I see thee face to face.[9]

Boulding was later to develop highly sophisticated analyses of peace and of methods of resolving conflict. It is important, however, to keep in mind the mystical basis of such work—and, indeed, of all of Boulding's concerns. Western thought is accustomed to drawing a sharp distinction between "mystical" and "rational" knowledge. The full-fledged mystic is ushered into a separate epistemological world and quietly left there. Those with mystical tendencies are expected to lead schizoid existences, segregating what they know according to the positivistic bias of the modern world. Yet this dichotomy remains open to question. Socrates, the rational analyst of Greek ethical and religious beliefs, was guided by an inner voice and subject to deep mystical trances. The twentieth-century Jesuit and mystic Pierre Teilhard de Chardin was a leading geologist and paleontologist. Such examples cast doubt on the necessity for a radical divergence between "inner illumination" and the rational pursuit of truth.

Teilhard, the contemporary thinker who has most influenced Boulding, wrote on the "mysticism of science." Science itself, he argued, embodied religious ideals:

> Rightly or wrongly, modern man has put his interest in an unbounded destiny beyond himself. And we have now all embarked to explore and conquer that future. Hope in a limitless future: the two essential characteristics of a religion.[10]

But without a higher mysticism of love, science would collapse in materialism, mechanization, brute force, and amorality:

> In order to sustain and extend the huge, invincible and legitimate effort of research in which the vital weight of human activity is at present engaged, a faith, a mysticism is necessary.

Whether it is a question of preserving the sacred hunger that impels man's efforts, or of giving him the altruism he needs for his increasingly indispensable collaboration with his fellows, religion is the soul biologically necessary for the future of science. Humanity is no longer imaginable without science. But no more is science possible without some religion to animate it.[11]

In this view, mystical and rational knowledge find harmony in the region of science's highest purpose and humanity's greatest good. Can such a relationship be found at less exalted levels? Out of his spiritual crisis, Boulding began his long struggle to bring the mystical and rational sides of his own life and work together.

At the spring session of the Syracuse quarterly Quaker meeting in 1941, he met Elise Marie Biørn-Hansen, a blond, Norwegian-American graduate student who was applying for membership in the Society of Friends. After a whirlwind courtship, the two were married in August.

Meanwhile, Boulding had decided to leave Colgate; the pleasant life in Hamilton had not been enough to offset four years as an instructor without a raise in his $2,400 salary. He accepted a job as an economist with the League of Nations Economic and Financial Section, which had recently moved to Princeton, New Jersey. His task was to study the problems of European agriculture. But his greatest concern was now the problem of World War II.

Having emerged from his spiritual crisis with his pacifist faith renewed, Boulding felt a deep responsibility to communicate to others the necessity of abolishing war through disarmament. For him the question of war was fundamentally religious. The proximate cause of war, he realized, was institutional: "We have war because there are *independent countries*; bodies of people organized for the essential purpose of maintaining their national independence by war." Yet the solution was not a world state; a sense of the need for diversity, plus a profound distrust of the centralization of power, prevented him from advocating such a course. A world state would hold a monopoly of all the means of violence, unchecked by other sovereign powers; a nation protected the area in which freedom was not a concept but a living reality. "It is right," said Boulding, "that men should have a home, and it is right that they should have a homeland. Our problem is to destroy the spirit within a nation that gives rise to war."[12]

The warlike spirit, he thought, could only be destroyed through the agency of a power which transcended the cares of this world. Such was the message Boulding gave in his William Penn Lecture of 1942, "The Practice of the Love of God."

> True forgiveness comes only in a flood of divine love, that wells up in our souls from places too deep to be hurt by mortal injury, love that draws us together with God and with our enemy in a healing, uniting experience.[13]

This was why, "apart from the love of God, there is no end to the cycle of war."

Boulding's strong convictions about war were shared by his new wife. Together, in the spring of 1942, they drafted a circular, "A Call to Disarm." Fellow Quakers considered it seditious and urged them not to distribute it. When Boulding showed it to his superiors at the League of Nations, he was told that if it were sent out he would be fired. He resigned his post, and copies were distributed that summer. The Bouldings were fearful of serious repercussions, but somewhat to their chagrin, the "seditious" document provoked not the slightest response. In the meantime, Kenneth accepted a job offer from Fisk University, a small black school in Nashville. Two years later, he and Elise issued a second pacifist letter, calling on all people to abandon their allegiances to earthly countries and join the Kingdom of Truth. This letter received as little reaction as the first.

After a year at Fisk, Boulding received a highly attractive offer to come to Ames, Iowa, as an associate professor at Iowa State College. There the economics department had pioneered in developing a multidisciplinary approach to agricultural problems, an approach which came to be called the "Ames School." Under the creative leadership of the department chairman, Theodore Schultz, the Ames approach was being extended to other areas of economics. To this end, Schultz wanted to find a general economist, give him a year to become familiar with labor issues, and thus create a labor economist uncommitted to the conventions of the field. Boulding was recommended for the job by Albert Hart, an old friend from Chicago. He happily accepted.

Despite good intentions and a year of study, Boulding never settled into labor economics; only one paper written at the time, "Collective Bargaining and Fiscal Policy," reflects his association with this specialized area.[14] Yet participating in the Ames School affected him profoundly. As he wrote later, "I did not really succeed in becoming a labor economist, but the effort ruined me as a pure economist. I became convinced that in any applied field one had to use all the social sciences. . . ."[15] At Ames he began for the first time to integrate his insights into economic theory with his convictions about the nature of man and the future of society. In 1945 he brought out his second book, *The Economics of Peace*, which attempted to outline the basic needs and principles of postwar economic reconstruction. The book had grown out of the work Boulding had done at Princeton, but reflected his new view of economics:

> Economic problems have no sharp edges; they shade off imperceptibly into politics, sociology, and ethics. Indeed, it is hardly an exaggeration to say that the ultimate answer to every economic problem lies in some other field.[16]

Kenneth E. Boulding

The analytical framework of *The Economics of Peace* was Boulding's "bathtub theorem," which held simply that the rate of accumulation was equal to the rate of production less the rate of consumption. Production could be likened to the flow of water "from the faucet, consumption to the flow down the drain. The difference between these two flows is the rate at which water in the bathtub—the total stockpile of all goods—is accumulating."[17] The principal economic effect of war was to drain the economic bathtub in a great waste of consumption; filling the tub again was thus the first task of reconstruction. The size of the tub was not, however, unlimited; a time would come, as stockpiles were increased, when the rate of accumulation would decline. This rise in the stockpile to the point where people were no longer willing to hold further accumulations was the fundamental cause of the end of a boom and a subsequent slump and unemployment, which Boulding saw as the most serious problem facing modern society. By its massive destruction of wealth, war put off the day of reckoning when mankind would have to face living with an economy that did not grow:

> It is the shadow of the classical "stationary state" that hovers over our day, and though it may be postponed by wars, by new discoveries, and by the opening up of new geographical areas to investment, yet these things only seem to be a postponement.[18]

It was with the problems of the stationary state, therefore, that postwar planners would ultimately have to grapple. The postwar boom would finally end, and the horrible danger would then be that another war would be needed to drain the tub. How could the familiar cycle of death and reconstruction be broken? Competition, despite its virtues in promoting economic progress, was insufficient to deal with the real problems of a stationary state. The modern peacetime economy would have to reflect a synthesis of capitalist and socialist elements.

The Economics of Peace was generally well received by its academic critics. That it never was a resounding success with the wider audience to which it was addressed may have had something to do with its moralizing tone and with its appearance at a time when most people were still thinking only about winning the war. The book was truly ahead of its time —and of later preoccupations with conservation and human welfare in a slow-growth or no-growth economy.

V

BOULDING believed that peace, economic stability, and the safeguarding of the natural environment would require mankind to turn away from greater and greater accumulation of capital goods. Consumption, he contended, was neither a trivial nor a merely technical issue; the very survival of society depended on a correct understanding of it.

> It is no exaggeration to say that consumption is the most important and intractable problem of a mature capitalism. As the total stock of real assets grows, the time must come when the rate of accumulation—i.e., of investment—must decline, and eventually indeed must decline to zero. . . . The great problem of the modern age, therefore, is how, eventually, to increase consumption to the point where full production can be maintained.[19]

Boulding eventually decided to give a full-scale treatment of economic analysis by means of his new theoretical apparatus. The result was *A Reconstruction of Economics*, which appeared in 1950. This work was his main contribution to economic theory; it sought, in effect, to explain economics in terms of capital rather than income. The concept of liquidity preference—the desire to hold cash rather than assets that earn a return and are less easy to convert into cash—enabled Boulding to see the purchase of goods and services not as a series of income flows to producers and suppliers but as a set of asset transfers. Here the basic analytical framework was the balance sheet. Liquidity preference also permitted Boulding to solve the problem of value, which had plagued earlier debates on the nature of capital. Capital had no intrinsic value; it was merely a heterogeneous collection of physical objects which were continuously being valued and revalued.

> Valuation is an essentially "present" process, arising out of the opinions, beliefs, and sentiments of the owners of physical capital (including money) operating on the stocks of the various forms of the physical capital actually in existence at the moment of valuation. Neither the past history of capital nor even its future prospects can affect its present value except in so far as these things affect the state of mind of its owners. It is true, of course, that both past history and future prospects do affect the present state of mind of owners, but this connection is by no means a stable one, and valuation always proceeds through the intermediary of the present state of mind, or "asset preferences," of owners.[20]

Boulding could thus revive his application of population theory to analyze the composition of capital stocks without becoming embroiled in the question of the effect of periods of production on the price structure.

A Reconstruction of Economics embodied a unified approach to both micro- and macroeconomic phenomena. Asset transfers, as expressed in a balance sheet, provided the best understanding of economic behavior, whether of the firm or of society at large. Boulding had, in fact, construed economics as a discipline for handling inventory. What mattered was what was "in the bathtub." Consumption meant the destruction of assets, calling forth production at the other end of the process and requiring the consumer to have an income. But enjoyment was in many cases different from consumption, since one enjoyed a painting by having it, and so too with a violin, a house, or a pair of ice skates (eating was an obvious exception). Maximizing welfare, enjoying one's assets to the utmost, meant minimizing consumption and, from a social standpoint, production and income. In modern capitalist societies, however, the minimum acceptable level of consumption was likely to be rather high, given the desire for full employment. But this could deplete the earth's resources and lead to less human welfare. Eventually, Boulding held, the rate of accumulation would decline to zero; with output just equaling consumption, we would attain the equilibrium of a stationary state.

No economist pursued the lines of thought which Boulding had advanced in *A Reconstruction of Economics*. The economics profession was just emerging from the upheaval of the Keynesian revolution and was not about to reconstruct itself again along Boulding's non- or anti-Keynesian lines. For his part, Boulding did not recant. But, though he continued to believe in the fundamental validity of his views, he did not press them.

If *A Reconstruction of Economics* did not point the future direction of economics, it did give some indication of the future course of Boulding's thought. In the preface he had written, "I have been gradually coming under the conviction, disturbing for a professional theorist, that there is no such thing as economics—there is only social science applied to economic problems. Indeed, there may not even be such a thing as social science—there may only be general science applied to the problems of society."[21] The book's first two chapters were titled "An Ecological Introduction" and "The Individual Social and Economic Organism"; they contained such subheads as "Reality as an ecosystem" and "The homeostasis of the balance sheet." Boulding's early interest in population theory had clearly been expanded into a broad, "organismic" view of society. At the heart of the bathtub theorem was, of course, the issue of conservation. Under the influence of Norbert Wiener's recently published *Cybernetics* (1948), he had come to see the problem of reaching economic equilibriums as part of the general theory of homeostatic or self-regulating mechanisms.

At Ames, Boulding had become convinced that to understand and solve economic problems, it was necessary to look beyond the conventional bounds of economics. He felt, moreover, that he was part of a growing movement of scholars seeking to unify the social sciences:

> We are, I believe, on the threshold of a new attempt at integration in the social sciences, perhaps even in general science. Past failures in this respect—notably those of the Spencerians, the Marxists, and the Institutionalists, who integrated a number of specialized errors into erroneous generalities, should not discourage us. The integration is clearly coming, not by the erection of a vast and tenuous superdiscipline, but by the reaching out of all the specialized disciplines towards one another.[22]

A Reconstruction of Economics was one attempt to move economics toward such an integration. It was to this great goal that Boulding would now increasingly dedicate his efforts.

V I

BOULDING'S stay at Iowa State was interrupted by a year (1946–1947) at McGill University, where he served as chairman of the economics department. In June 1947, John Russell, the first of five Boulding children, was born. The family returned to Ames for the fall semester, with Boulding now established as a full professor. While in Canada, he had flirted with the idea of becoming a Canadian citizen; he now decided, as had long been his intention, to seek United States citizenship. He had taken out the necessary papers when he was at Colgate, but was fearful that final approval would be withheld because of his refusal to take the oath to bear arms. (During World War II, Boulding had not applied for conscientious objector status, since he was unwilling to perform even the alternative service required of pacifists; drafted in 1944, he escaped going to jail when an Army psychiatrist, unable to come to terms with his description of the Quaker notion of divine guidance, awarded him a 4-F classification.) Boulding was prepared to carry his case for citizenship all the way to the Supreme Court. This turned out not to be necessary, for a sympathetic lower court judge found in his favor. In December 1948, Kenneth Boulding became a United States citizen.

In 1949, the newly established American became firmly entrenched in the front rank of his chosen profession. He left Ames to become professor of economics at the University of Michigan in Ann Arbor, which was considered to have one of the nation's top economics departments. At the annual meeting of the American Economic Association in December, he received the John Bates Clark medal, given every other year to "that American economist under the age of forty who is adjudged to have made a significant contribution to economic thought and knowledge."

Boulding was now a name to be conjured with. An increased demand for his work brought forth a literary output of remarkable proportions. He had always been industrious; between 1932 and 1949, he had published twenty-seven articles, six book reviews, three pamphlets, two books, and a collection of religious sonnets based on the dying words of a Quaker martyr. But from 1950 to 1972, his output included some 300 articles, more than 100 book reviews, eleven pamphlets, nine books (excluding various new editions, collections of his own articles, and edited collections of others' articles), and miscellaneous verse.

Boulding considered that he had become a member of several different "invisible colleges"—a term he picked up from a Yale historian of science, Derek de Solla Price, who was intrigued with the discovery that Invisible College had been the original name of the British Royal Society. In *Science Since Babylon*, Price advanced the idea of new invisible colleges consisting of groups of scientists bound together by a common interest in a particular area of research and communicating by means of personal contact and informally circulated papers.[23] Boulding found the idea congenial. For him, the idea of an invisible college had a spiritual dimension; in it, people were unified not by worldly ties but by a common intellectual purpose. As such, the "intellectual college" was the perfect instrument of what Teilhard de Chardin called the noosphere, the earth's envelope of thought and knowledge which both liberated and bound together all mankind.[24]

In economics, Boulding had never felt part of an invisible college; perhaps this was because he had worked largely on his own in the area of economic theory, reaching rather idiosyncratic conclusions. He was neither Keynesian nor Chicagoan. His first invisible college in economics was the "Ames School" at Iowa. At Michigan he began to move toward a broader invisible college in general systems; even his contract called for him to pursue interdisciplinary studies. During his first years at Michigan, he conducted annual faculty seminars dealing with the integration of the social sciences. Each participant at these seminars was expected to present a paper from the standpoint of his own discipline on such topics as "The Theory of the Individual," "Growth," "Information and Communication," and "Conflict." Boulding's participation is reflected in a series of papers, including "A Conceptual Framework for Social Science,"[25] "Economics as a Social Science,"[26] "Toward a General Theory of Growth,"[27] "Contributions of Economics to the Theory of Conflict,"[28] "The Malthusian Model as a General System,"[29] and "Notes on the Information Concept."[30] The major work of Boulding's early career at Michigan was, however, *The Organizational Revolution*.

In 1952, Boulding had been commissioned by the then Federal Council of Churches (later merged into the National Council of Churches) to make a study of the ethical implications of the society's increasingly large economic organizations. *The Organizational Revolution*, which appeared in 1953, combined a broad, interdisciplinary approach to economic institutions with a deeply felt expression of religious morality. Boulding was thus able both to urge his case for a unified social science and to pursue a recent interest in the relationship between economics and religion. This new interest had found expression in a number of papers written at this time: "Protestantism's Lost Economic Gospel,"[31] "Religious Perspectives of College Teaching in Economics,"[32] "Religious Foundations of Economic Progress,"[33] and "The Quaker Approach in Economic Life."[34]

In the preface to *The Organizational Revolution*, Boulding set forth two points for the reader to keep in mind. First, he did not intend to eschew the role of moralist. "In this study . . . I shall write unashamedly in part as a moralist. I will, however, endeavor to make the moral system as explicit as possible."[35] Second, Boulding saw the organization as a kind of organism, possessing a unity susceptible of coherent description, characteristic modes of behavior, and a certain machinery of existence. As an organism, the organization had to be considered in the context of its ecological surroundings. The totality of human organizations, like the totality of biological organisms, constituted an ecosystem, a self-contained and self-perpetuating system of interacting populations.

But Boulding did not make the mistake of suggesting that human organizations are literally the same as biological organisms; on the contrary, he stressed that the constituent parts of organizations were conscious beings with wills of their own. Hence, organizations must deal with the problem of consent. Boulding concluded his preface by outlining what he considered to be the most serious problem which organizations present to society:

> The necessity for hierarchy in the structure of organization has created a severe moral dilemma which is by no means yet resolved. On the one hand we have the pull toward the organizational necessities of hierarchy, toward an aristocratic, highly stratified society of status. On the other hand there is a profound pull toward the moral ideal of equality—a pull which is especially strong in societies which have been affected by Christianity, with its emphasis on the equality of all men before God and on the universality of love.[36]

Boulding began by setting forth some basic characteristics of organizations. They tended to be benign toward insiders, hostile toward outsiders; they satisfied man's need for status by formalizing his place in society and thus making him feel more secure. The expansion of organizations was limited both by an increasingly unfavorable environment and by an increasingly unfavorable internal structure. This view of the limits of organizational expansion set up the book's principal thesis. The growth of economic organizations since about 1870, Boulding argued, had been the result of certain technical improvements in transportation, in communications, and in organizing skill. These improvements had lessened the internal resistances to growth and made the external ones more important. Economic organizations could and did grow to alarming size—this was the revolution—and therefore had provoked such governmental restrictions as antitrust legislation. In fact, Boulding claimed, "one can almost describe the history of the present era as a continuous encroachment of politics on economics."[37]

In economic terms, the advent of the new organizations led to both an increase in productivity and an impairment of the free market. Less and less were the prices of goods and services determined impersonally through the participation of individuals and numerous small firms in the market; instead, they had become increasingly responsive to the decisions and preferences of large labor unions and oligopolistic corporations. Increases in prices and wages were more acceptable to these organizations than were decreases; the modern economy thus acquired an inflationary bias. His thought here paralleled Galbraith's. Boulding was, however, most interested in the ethical implications of the organizational revolution. These he took to be wrapped up with the very essence of organization.

> The relation of ethics to organization can be summed up in the question *how* wrongs are righted; what machinery exists in the world for the correction of conditions which are perceived to need correction. This machinery, however, is precisely what is meant by organization.[38]

For Boulding, the organization was basically a system of communication, hierarchically arranged, in which information was passed in increasingly refined form from lower levels up to a decision-making executive. For the organization to survive, the executive needed an accurate picture of the outside world on which to base his decisions, i.e., to "right wrongs." The larger the organization, the more difficult was the task of putting together such a picture. Increasingly, therefore, the organization had come to depend on the efficient coordination of its personnel. Boulding saw a threat to human liberty in this subordination of men to the needs of the organization.

Boulding's paradigm for an effective organization was the thermostat. The thermostat was the homeostatic mechanism

par excellence, designed to correct automatically any devia-
tion from an "ideal" temperature. What Boulding really
wanted to know about organizations was why they do not
right wrongs. His answer was twofold. First, there might be *II*
technical defects in the organizational structure: gaps in
communication, inadequacies of the executive, mistakes of
the lower-echelon people. Second and more important were
the moral issues. Organizations might themselves be designed
to do things which were not right, or the ideal values of the
variables which govern the organizations' behavior might be
wrong values. With these two concerns in mind, he turned to
a consideration of several types of modern organizations: the *II*
labor movement, the farm organization movement, business
organizations, and the national state.

In each case, Boulding emphasized the ascendance of personal and political relations over exchange in economic life.
Much of the strength of a large organization lay in its ability
to provide its members with desired status and to elicit moral
and emotional support from them. At the same time, it possessed and exercised the power to coerce both its members
and outsiders to further its own ends. Boulding thus began to
see the economy—and, indeed, society as a whole—as subject to three organizing forces: the exchange of goods and
services for mutual benefit, coercion or fear of reprisal, and
love or the integration of desires and objectives. Unfortunately, the chief alternative to exchange seemed to be fear:

> As organizations grow larger and larger, relationships must of
> necessity become more and more formalized, and the most
> acute problem of society is to achieve the right degree and
> kind of formalization. A society whose theoretical structure
> has never faced this problem, and which tries to apply a fami-
> listic ethic to a brontosaurian organization, will end in a ter-
> roristic rigidity. In our present state of knowledge the only
> substitute for the cash nexus is the fear nexus: a society
> moved not by the hope of gain but by fear of the inquisitor.[39]

Organizations were here to stay, but they did not augur well for the future of man. They had forced free-market exchange into the background. Did that mean that the inevitable destiny of humanity was a society based on fear? Boulding saw but one way out of this dilemma:

> We have reached the pass now in the development of man where there is almost literally no choice but love; where the only basis of human organization which can function is that of common concern and common need. It is on this slender foundation that the future of man must be built.[40]

A doctrine of love lay at the heart of Boulding's religious belief. A decade earlier, he had asked a Quaker audience how it responded to the words "God is Love." "Perhaps," he suggested, "they lead you into a comfortable corner of your soul, well insulated from the chilly world of rational thought, where you secretly indulge in spiritual drinking." This and similar responses signifying less than total commitment were unfortunate:

> If any of these conditions is yours, then you have missed a treasure. For to some these words are a key to a Kingdom, a Kingdom where Truth reigns in so great majesty that we can hardly bear the splendour, where life springs born again from every moment of time, and where a rich joy compounded of bitter spices scents every breath we breathe.[41]

In the *Organizational Revolution*, Boulding tried to give love a respectable place in the chilly world of social theory. He did not accept the view that two opposite coercions would cancel each other out, as Galbraith optimistically held in his doctrine of "countervailing power." The danger of counter-coercion was only too apparent, Boulding felt, in "the appalling breakdown of national defense in our day."[42] Rather, competition in love had to be substituted for competition in fear. By this, Boulding meant that a moral and organizational environment had to be created in which "those organizations

Kenneth E. Boulding

which are not meeting the needs of man, and which are not serving to right wrongs, will not survive in competition with those organizations which are meeting the needs of man."[43]

The social-democratic state had the capacity to establish such an environment within its borders. It possessed mechanisms for the righting of wrongs; it had rules governing the behavior of organizations, including itself; it could combine a healthy plurality of interests and institutions with a sense of common purpose and responsibility. Unfortunately, such a sense of common purpose was developed to no small extent through wars with other nations. And here lay the greatest danger of all:

> It is the complete failure of the social-democratic state to solve the problem of its defense which threatens to suppress all its other virtues and to transform it into something different —a militarized garrison state—or to overthrow it altogether.[44]

Boulding offered no program for eliminating war. Distrusting the immense coercive powers which a world state would have, he also suspected that it would prove unable either to attract the support or solve the problems of the entire earth's population. Rather, he stressed the importance of ethical vision: international cooperation must proceed at all levels based on the principle of universal brotherhood. For there was no simple mechanical solution to the problem of organizations.

> The final conclusion, therefore, is that though organizations are here to stay and though the only solution to many of the problems which they raise seems to be ever more and larger organizations, yet there is no substitute for the Word of God —the sharp sword of truth in the prophetic individual, the penetrating moral insight that cuts through the shows and excuses of even the best-organized society.[45]

[221]

Included in the first edition of *The Organizational Revolution* was a series of basically friendly critiques by a number of economists, sociologists, and theologians. One of the contributors, Reinhold Niebuhr, disagreed with Boulding's fundamental conviction that the great evil confronting society was coercion. Rather, Niebuhr argued, it was man's inherent selfishness. In his response, Boulding noted that this disagreement reflected a difference in theological positions. He saw man as essentially good, though corrupted by outside constraints, whereas Niebuhr saw him as inescapably sinful. Boulding here expressed the Quaker notion of man's perfectibility, which in fact underlies all his work. Knowledge, the original cause of man's Fall, now becomes the agent of his salvation, freeing him from the trammels of malignant external forces and allowing his natural goodness to emerge. The task of increasing knowledge is thus an essential part of God's work.

VII

IN 1954, Boulding went to Palo Alto, California, as a fellow of Stanford University's newly opened Center for Advanced Study of the Behavioral Sciences. Among those also present during that inaugural year were anthropologist Clyde Kluckhohn, political scientist Harold Lasswell, and the biologists Anatol Rapoport and Ludwig von Bertalanffy, the latter the father of general systems theory. All were interested in the problem of developing a general theory of behavior out of the various social and biological sciences. Together they founded the Society for General Systems Research. Boulding was appointed president, and Rapoport and von Bertalanffy were designated as coeditors of a *General Systems Yearbook*, which appeared for the first time in 1956.

During Boulding's year in Palo Alto, a new invisible college in peace research also emerged. One of the junior fellows at the Center for Advanced Study was Stephen Richardson, whose late father, Lewis Richardson, had done pioneering (but largely ignored) work quantifying historical data and building mathematical models of the conduct of wars and arms races. The young Richardson generated considerable excitement by making available microfilm copies of his father's books, since war was obviously the greatest behavioral problem which society faced. At the same time, another junior fellow, the social psychologist Herbert Kelman, asked the people at the Center for advice on how to improve *Research Exchange on Prevention of War*, a newsletter he had been distributing for several years.

Boulding was greatly impressed by Lewis Richardson's work; he had increasingly come to feel that war, like other human problems, could only be eliminated through a precise understanding of how and why it came about. Kelman's query suggested a vehicle for gaining and communicating such an understanding. Together with Rapoport, who was about to come to Ann Arbor, Boulding recommended that the newsletter be expanded and that it originate at the University of Michigan. A list of sponsors was put together, and the proposed publication was given a new title, *Journal of Conflict Resolution*. Boulding became chairman of the editorial board, and in 1957 the first edition appeared under the auspices of the Michigan School of Journalism.

Two years later, Boulding was able to obtain permission to establish the interdepartmental Center for Research on Conflict Resolution, and the *Journal* became its official publication. Boulding himself contributed such articles as "Organization and Conflict"[46] and "National Images and the International System."[47] Never retreating from pacifism, he was nonetheless convinced that moral outrage and a principled opposition to war were no substitute for knowledge and understanding. Even at the height of the Vietnam war, Boulding defended peace research against the attacks of fellow Quakers who felt that the cause of peace was best served not by research but by proselytizing and demonstrating.

Near the end of his year in Palo Alto, Boulding, in a state of intense intellectual excitement, spent just nine days dictating the whole of a new book, *The Image*. This was his attempt to provide a general epistemology for dealing with all forms of behavior. All knowledge, in Boulding's view, is constituted in a series of images—of the world and the knower's position in it, of the past and the present, of values and desired objectives. These images are revised as new information is added to them; indeed, the meaning of a new bit of information *is* the change it produces in an image. Obviously, anything perceived, sensed, or imagined can help to build an image. Coherent images are not the product of just one sort of data. There are strictly personal images, and ones which are shared by many people. In all cases, however, knowledge is arrived at in the same fundamental way.

But which images provide a true picture of independent, underlying reality? How can the degree of correspondence between image and reality be ascertained? Boulding deliberately avoids this problem, saying that he has tried to steer clear of the great philosophical issue of whether an image is "true," or how, if it is true, we know that it is true. Rather, his purpose is to "discuss the growth of images, both private and public, in individuals, in organizations, in society at large, and even with some trepidation, among the lower forms of life."[48] His object is not greater philosophical precision but a more adequate means of understanding behavior.

Postulating the image as the basic unit of knowledge, his study begins by outlining seven levels of organization: organizations with static structures, clockwork organizations, organizations with homeostatic control mechanisms (where the concept of information, and hence the image, first becomes important), cells, plants, animals, and humans. At each successive level, Boulding argues, images grow more sophisticated and complex, while the concept of the image becomes increasingly important for any theoretical model of behavior. Man, with his unique characteristic of self-consciousness, is seen as acting on the basis of images not only of his environment but also of himself. These images are amalgams of fact and value held by varying numbers of people and susceptible in varying degrees to change. The great desideratum is to gain better images, and thus to improve both the efficacy and the moral caliber of our actions. Boulding believes that the ultimate goals of men and of societies as a whole are fundamentally good and fundamentally the same, although bitter disagreements arise over the means for achieving them. Improved images would point the way more clearly.

After a survey of the place of the image in biology, psychology, sociology, economics, and history, Boulding proposes a new field, to be called *eiconics*, which would study the nature and growth of images generally. In effect, eiconics would provide the image of the image. Based on Gestalt psychology and the theory of feedback developed in cybernetics, eiconics would be a truly interdisciplinary field encompassing all the behavioral sciences, ranging from biology to political science. In essence, it would provide a broad theory of the formation of knowledge and its transformation into action. Knowledge for Boulding is the main hope of mankind, for it is anti-entropic; that is, unlike physical processes, it does not tend toward a random or run-down state. Unlike energy, knowledge can be increased. It is lessened neither by use nor by transfer from one person to another.

Boulding did not see himself as the definitive developer of his new science of eiconics:

It may be that like Moses I have only brought the reader to Nebo, from which tantalizing glimpses of a promised land may be obtained. Like Moses, also (and let the comparison stop there), I am pretty sure that I shall not go over Jordan. Some Joshua must arise if the promised land is to be taken.[49]

Although *The Image* received some very favorable reviews, it called forth no Joshua to lead the people of academe into the promised land. In the course of time, Boulding found that three sorts of people were most receptive to the book: teachers of the liberal arts in engineering schools, Madison Avenue-type executives, and those interested in decision theory, management science, and general systems. By and large, however, Boulding had been wrong in prophesying a great new push toward the integration of the sciences. New fields like biophysics or urban studies, combining the characteristics of a number of disciplines, might arise where there was a strong need to treat a particular set of phenomena which cut across disciplinary lines. But most scholars were not prepared to abandon their bailiwicks to chase what they considered to be the chimera of a unified science. There was enough to do as it was. Thus, *The Image* and the new science Boulding hoped to spark came to languish in yet another small subculture of academic society. Nevertheless, Boulding did not allow the concept of the image to slip into oblivion; rather, he used it in much of his subsequent work as a way of analyzing the behavior of men and institutions.

VIII

BY THE TIME he returned to Michigan after the year in Palo Alto, Boulding's work looked more and more like a juggling act in a circus of the intellect. Economics, general systems, the study of conflict resolution and peace, social theory, philosophy of history, religion—books and articles of his appeared in all these areas. In the 1960s he joined two more invisible colleges: ecology, a suddenly popular subject in which he had long been interested; and grants economics, a field which he founded dealing with one-way transfers of goods, services, or money.

Despite his incredible variety of subject matter, Boulding's work during this period was nothing if not cohesive. This can be seen through a consideration of Boulding's critique and defense of economics. Boulding's main reproach to economics was that it lacked sufficient scope. His experience at Ames had convinced him that there was a great need for a unified social scientific approach to economic phenomena.

Boulding thus came to have a new respect for the American institutionalist economists Thorstein Veblen, John R. Commons, and Wesley C. Mitchell, who had attempted just such an approach. In a 1957 article, "A New Look at Institutionalism," he went so far as to place himself directly in their camp:

> In a letter to me a few months ago, Professor Ayres accused me of having become an institutionalist. If a somewhat despairing concern for dynamics in theory (without losing a sense of the very real accomplishments of statics); if a very strong concern for integration in the social sciences and for the bringing of contributions from psychology, sociology, and the biological sciences into the construction of better theories of individual behavior and social change; if a strong (if sceptical) interest and sympathy with empirical methods is enough to make me an institutionalist, then I gladly accept the title.[50]

Boulding was not a socialist; he believed in a pluralistic capitalism based on a market economy. He was, however, sharply critical of the economic libertarianism which was expressed most forcibly by Milton Friedman. Interestingly enough, Boulding did not deny the value of the market as an agent of resource distribution and a mechanism for economic decision making. Rather, his opposition to Friedman was based on the conviction that a large part of economic activity and the motivation for it lay necessarily outside the reach of the market and of market forces. He framed his argument in terms of the three organizers of society—love, fear, and exchange—which he had set forth in *The Image*. Economic libertarianism, he argued, did not look beyond exchange; it failed to take into account the vast amount of economic activity predicated on love (the "integrative system") and fear (the "threat system").[51] Taxation, an institution of legitimized coercion, was part of the threat system. The upkeep of one's children and aged relatives, foundation grants, and government subsidies, pensions, relief payments, and aid to dependent children were all part of the integrative system. All three organizers, Boulding held, were necessary in any society. An exchange economy could not operate successfully without both a broad integrative matrix of trust and allegiance and the legitimized threat system represented by a code of law. To look exclusively at exchange meant misrepresenting the general nature of economic life, as well as ignoring much economic activity. Boulding did not dismiss the importance of exchange as a social organizer; what he wanted was a more comprehensive economics.

Boulding termed as the grants economy that part of economic life which did not involve exchange. This notion was first set forth in "Notes on a Theory of Philanthropy" (1969).[53] He himself received a small grant from the Ford (1962),[52] and elaborated in "The Grants Economy" Foundation for the study of grants, and together with an enthusiastic associate, Martin Pfaff, Pfaff's wife Anita, and Janos Horvath, he established the Association for the Study of the Grants Economy. In 1973, Boulding published *The Economy of Love and Fear: A Preface to Grants Economics*, which argued the general proposition that "the one-way transfer, far from being something extraneous or extraordinary in the general organization of social life, is an integral and essential part of the system, without which not only community but organization and society itself would be virtually impossible."[54] In this book he developed a micro and a macro theory of grants and applied both theories to charity phenomena, public finance and the provision of public goods, the theory of organization, and various contemporary political controversies.

In a series of lectures given in 1953, Boulding had distinguished between the economic ethic and the heroic ethic in man. The economic ethic was expressed by the merchants and traders who relied on the "lore of nicely calculated less and more"; the heroic ethic was the possession of civilization's great saints, sinners, and standard-bearers. Both ethics, he felt, were needed in society:

> . . . there is a great deal of *complementarity* between the economic and the heroic in the last great production function of the universe; this is to say that attempts to be purely economic or purely heroic are generally unsuccessful. Without the heroic, man has no meaning; without the economic he has no sense.[55]

Two decades later, Boulding was able to find in the study of grants a way for economics to approach the heroic side of man, where the deepest motives were love and fear, not the desire to turn a profit. Economics would thus become at once more true to life and more morally sensitive. What better way to broaden its scope!

If Boulding criticized economics for being too narrow, he also proselytized for the broader application of ecnomic techniques. *Conflict and Defense* (1962),[56] his major contribution to peace research, relied on the theory of oligopoly as the primary means of analyzing various forms of conflict. Boulding combined the economic analysis of imperfect competition with Lewis Richardson's models of international conflict and the game theory of John von Neumann and Oskar Morgenstern, in an attempt to develop a general theory of conflict. His object was to demonstrate that conflict processes were not arbitrary, random, or incomprehensible, for he firmly believed that in the understanding of those processes lay the opportunity for their control. Economic theory provided tools for explaining behavior which was not in large part economic. In *Economics as a Science* (1970), Boulding discussed the place of economics in the general study of society. He suggested that it might be valuable to have a notion of "biological profit" and to discuss ecological equilibrium in terms of the economic conception of the invisible hand.[57]

Perhaps most significantly, Boulding emphasized the importance of economics in providing a general theory of value. He was convinced that one of the major weaknesses of modern thought was a positivistic bias which treated values and value judgments as givens not susceptible to rational analysis. Boulding's desire for a broad, reasoned discussion of values was bound up with his deep religious feelings. In "Some Contributions of Economics to Theology and Religion," the discussion centered on the question of value:

> If the economist, then, has anything special to say in the area of the Great Values it is that values, whether great or small, are always the result of acts of choice. Values do not "exist" independently of the actions of the valuer; they are quantities which are conveniently descriptive of acts of evaluation. It is these acts which exist, not the values. The "price" of a commodity is not a physical quantity like its weight; it is a quantity descriptive of an act, either of exchange or evaluation, and is meaningless without the actor. The economist therefore shifts the discussion of value from the value as a "thing" to the evaluation as an act.[58]

For Boulding, religious values too were inextricably bound up with the actions of people in this world. What mattered was what was done; this was the touchstone of religion.

Kenneth E. Boulding

Boulding believed, however, that the course of action which was selected was more likely to depend on the opportunities available than on abstract preferences, since the same set of basic preferences could produce widely differing behavior as the opportunities changed. On this proposition rested his best hopes for the world. By nature people were good; the object was to provide them with opportunities to exhibit this basic goodness. The creation of such opportunities depended, above all, on the expansion of knowledge.

From this brief survey of Boulding's later work, it is evident how naturally what he had to say in one area fed into what he had to say in another. Throughout these writings ran the same overriding ethical concerns. Embracing the whole was a broad ecological view of all aspects of life. Growth and development were fundamentally evolutionary, taking place gradually as populations shifted in size and nature. In Boulding's ecological model, knowledge, as represented by the image, played an important role, increasing in importance as the members of a population grew more sophisticated. At the highest level of man and society, change could take place as a result of conscious intent:

> With the development of self-consciousness in man, the evolutionary process became at least in part teleological, that is, directed by an image of the future in the minds of the active participants who were capable of affecting the system.[59]

Boulding elaborated his evolutionary view of human history in two books, *The Meaning of the Twentieth Century* and *A Primer on Social Dynamics*. The former was subtitled "The Great Transition"; it argued that in the twentieth century, society was undergoing a fundamental change from a "civilized" to a "postcivilized" state—that is, to a very high state of technological development. Boulding postulated three great ages of man. The first was one of precivilization, where people lived at subsistence level by hunting and gathering food. Between 8000 and 3000 B.C. a major transition began to take place which, by a two-stage process, created civilized society. There was, first, the invention of agriculture, which freed some fraction of the population from the necessity of producing food. This led in turn to the rise of cities. Under civilization, which lasted for perhaps 5,000 years, between 75 and 80 percent of all people were farmers, supporting the remainder by the surplus food they conveyed through some form of physical or spiritual (priestly) coercion. Sometime in the Middle Ages, all this began to change under the impact of technology. While the population itself increased dramatically, the percentage of the population needed for food production gradually decreased. An exchange economy began to supersede the old coercive system of economic exploitation. The standard of living rose.

Boulding argued that this second great transition from a civilized to a postcivilized world did not really begin asserting itself until the twentieth century. A world of prosperity, justice, and personal fulfillment was within reach. Progress toward this happy state of postcivilization was not, however, inevitable. War, overpopulation, and the exhaustion of natural resources were traps which could destroy society and which society had consciously to avoid. The successful negotiation of the great transition depended on people with an enlightened image of the future and an awareness of the dangers which lurked along the way. For Boulding, such people included Pierre Teilhard de Chardin, Aldous Huxley, and H. G. Wells; together they made up a small invisible college capable of shaping man's best destiny. For this invisible college, said Boulding, he was an unashamed propagandist.

The underlying thesis of *The Meaning of the Twentieth Century* was that the development of human society was an evolutionary process resulting from the growth of knowledge. The discovery of agriculture, the invention of new forms of social organization, the development of technology—these were responsible for moving man from pre- to postcivilization. Boulding saw this view of history as fundamentally opposed to the dialectical interpretation expressed in Marx's theory of historical materialism. Marx looked at history as a series of economic systems, each of which developed as the result of a violent confrontation between opposing interests inherent in its predecessor. Thus an original, primitive communism was supposed to have given rise to a slave system, which eventually led to feudalism, which in turn was superseded by capitalism. Marx's analysis—and attack—was focused on the capitalist system; under capitalism, Marx predicted, the plight of workers would become so acute that they would of necessity rise up and overthrow the dominant class of property owners, establishing a communist society once and for all.

Boulding not only disagreed with Marx's prediction but also opposed the notion that dialectical conflict was a significant and positive force in human history. His *Primer on Social Dynamics* was an attack on Marxian dialectics. Boulding contended that dialectical elements in society were principally contained in the "threat system" but were relatively unimportant in exchange and played almost no role in integration. All three of these social organizers were, in fact, aspects of the basic learning process which underlay all human history. The dialectical process did not advance the growth of knowledge but was a hindrance to it, since dialectical struggles prevented men from validating knowledge by direct or scientific methods and represented "a backward step towards the testing of images by the success of whole cultures, not by the testing of knowledge itself."[60]

With Boulding's conviction that the growth of knowledge was the crucial factor in pushing society forward, it is hardly surprising that he reduced revolutionary change to the status of mere incidents imposed on the larger pattern of development. What really mattered was the sum of human knowledge, or noosphere. Only through an expansion of the noosphere would man be able to shape his own destiny and make the human adventure truly teleological. Boulding's vision is more sweeping, abstract, and religious than, say, John Kenneth Galbraith's essentially political view of the role of knowledge and its developers, but they are talking about the same thing: the industrial system has brought into existence, to serve its intellectual and scientific needs, the learned community that may reject "the system's" monopoly of social purpose. In an age in which technology has shattered old social structures, wreaked terrible destruction through wars, and threatened the global environment itself, faith in knowledge as the source of salvation remains the common hope of the academic community.

Boulding's particular contribution has been to invest modern economic and scientific knowledge with moral content. He has sought not only to unify science and religion but also to embody both in his personal and public life. Since World War II, the antiwar movement has been a primary focus of his public activity. In April 1958 he initiated a vigil in Ann Arbor to protest the continued testing of nuclear weapons. Two years later, he participated with a thousand other Quakers in a peace vigil in front of the Pentagon. He opposed the Vietnam war from the beginning, and in 1965 he helped plan and conduct the first antiwar teach-in. In 1966, Boulding served as a delegate to the World Council of Churches' World Church Conference in Geneva, where he was lionized as a leading economist who was also a practicing Christian.

Boulding was named president of the American Economic Association for the year 1968. This turned out to be a more difficult task than usual, for the Association's annual convention had been scheduled to take place in Chicago. However, in the wake of the tumultuous Democratic National Convention and the police assault on the anti-war demonstrators, many economists wanted the locale changed. There was a lively controversy, but it was Boulding's decision to make. After much soul-searching, he surprised many of his colleagues by deciding to stick to the original plan, arguing that it was not in accord with the traditions of the Association to take unilateral political actions of this sort.

In 1967 he took a sabbatical from Michigan to be visiting professor of economics at the University of Colorado and director of the Program of Research on General Social and Economic Dynamics at Colorado's new Institute of Behavioral Science. He liked Boulder, liked the mountains, and was somewhat unhappy about the lack of support he felt he was getting from the University of Michigan for his various general systems and peace research projects. The following year he became a permanent member of the faculty of the University of Colorado, where he felt quite content to get on with his own work in his own way.

I X

WITHIN the economics profession, Boulding is regarded as something of a heretic, though a basically harmless one. As a person he is regarded warmly by his fellow economists; hats are tipped to him as one trying to expand and ennoble the profession. But as an economist per se, in terms of the contemporary criteria of the economics profession, he is taken less than entirely seriously. Such condescension is unlikely to last. His early work on price theory and the theory of the firm is too good to remain neglected forever. His stress on ethical values and his work on the theory of organization, general systems, and the nature of conflict have indeed enriched economics. His work on grants economics—not the happiest name for what he is talking about—has opened up a fertile

new field for an age in which the role of the market appears
to be shrinking.

Though Boulding has worked most of his life on the bor-
derline of economics, the price he has paid for his seeming
dilettantism has not been great. He has had as successful an
academic career as anyone might wish, doing the work he
wanted to do, publishing whatever issued from his volumi-
nous pen, serving as president of his discipline's most prestigious organization. Referring to his pacifist activities, Boulding has said that there were times when he made decisions of
principle, "but I am sure I'm not going to get much of a
reward in heaven, because really—every time I've made a
noble sacrifice, I've always gotten a better job." His whole life
is, in fact, an inspiring story of heresy tolerated, creativity
appreciated, and virtue rewarded.

By accepting the guidance of his religion, Boulding has in
fact focused his work as an economist on what is doubtless
the most crucial unsolved problem of our time—war, mankind's historic pseudo-cure for economic and social ills. Inevitably, in the midst of ecological threats, contests for access
to the world's limited resources, rampant inflation, growing
unemployment, famine in the poor lands, and a threatened
breakdown in the world monetary system, fears of war are
again on the rise. "Is it not tragic," asks Mubashir Hasan, the
Pakistani finance minister, "that it is mainly when nations
have plunged themselves into wars that there has been a sud-
den spurt of creative activity on the production front? Men
and women have toiled day and night, instead of working five
days a week, and achieved technological breakthroughs
which have tremendously boosted production and helped
usher in new eras of productivity later on."[61] Is Boulding's
"bathtub" about to be drained once more?

And is this mankind's ghastly little secret—that we really
need war to provide the spur for creative activity and social
cohesion, to inspire and develop all that is "best" in us? Other

scholars—and not just Karl Marx—have long thought so. For instance, Everett E. Hagen of the Massachusetts Institute of Technology, in his studies of Japan, concluded that military threats to a nation may be a powerful force toward economic growth, if combined with internal forces pushing toward technological development. This is no new idea. The ancients put great stress on the threats of enemies in developing the genius of a nation. The tragic irony is still with us. The most dramatically successful of all research undertakings in this century was the Manhattan Project, which enlisted and coordinated the talents of many brilliant scientists in the American effort to develop the atomic bomb before the Nazis did. On the other hand, the great wartime innovations owed their origin to major scientific progress in physics, chemistry, engineering, the earth sciences, the biological sciences, and mathematics which predated the war by decades.[62] The Einsteins, Lawrences, Fermis, and von Neumanns did not need war to inspire their search for knowledge.

Nevertheless, it is true that nations were energized by World War II, and fruitful human activity followed it. International organizations were created to expand trade, to maintain employment and real income, to correct maladjustments in the balance of payments, to reconstruct and develop economies on a more equitable and just basis. The war opened up an era of growth; but now the postwar era appears to have exhausted itself and is subtly, horribly, being converted into what feels like a prewar era. Is mankind preparing itself again for one of its dark nights of creation?

In the nuclear age—God forbid.

As the Pakistani minister put it,

the question before the world today is whether the destruction and bloodshed that a global conflict involves is really a prerequisite to a spurt in productive activity by nations and to a restructuring of the international economic order on a just basis. Must the nations first make the terrible sacrifice of human life and property before they bring themselves to taking those hard decisions which alone would save mankind from starvation and death?[63]

But this time the world cannot go through its historic blood rites of spring as a prelude to a new era of growth. The dangers to humankind and the ecosystem are too great. The nations must instead settle down to the hard work of finding a way to share the earth's limited resources equitably, to distribute income and capital and knowledge more fairly. This is not simply a contest between the rich nations and the poor, as the oversimplified rhetoric of the earlier postwar period had it, and still has it. Rather, the time has come to recognize that there can be injustice not only between rich and poor but between oil-rich and oil-poor, between those with ample food and those with none, and that such injustice exists within states and not just between states. Many of the poor states neglect their own poor most cruelly.

Kenneth Boulding has sought to teach us that we must learn to live together on the earth. He links the scientific and technological present to a sweeter and simpler, more humane and more divine past. But we are held down by our human past, by our mean antagonisms, by the rigidity of our habits and the faintness of our hopes.

In the Medieval miracle play "The Creation of Adam and Eve," God says:

> *In earth are trees and grass to spring;*
> *Beasts and fowls both great and small,*
> *Fishes in flood, all other thing,*
> *Thrive and have my blessing, all.*

> *This work is wrought now at my will,*
> *But yet can I here no beast see*
> *That accords by kindly skill,*
> *And for my work might worship me.*

Kenneth E. Boulding

For perfect work ne were there none
But aught were made that might it yeme;
For lof made I this world alone,
Therefore my lof shall in it seem.[64]

The last verse means, "For there would be no perfect work unless something were made that could have charge of it; for love alone I made this world, therefore my love shall be manifest in it."

Wildly, daringly, perhaps madly but most kindly, Boulding has given us the economics of the perfectibility of man and of God's earthly kingdom.

E Pluribus Unum

Do I know what rice is?
Do I know who knows it!
I don't know what rice is,
I only know its price.

Do I know what man is?
Do I know who knows it!
I don't know what man is,
I only know his price.
                        —*Bertolt Brecht*

I

ECONOMISTS have rarely been popular with the generality of people. This is strange, because economists have long insisted that their subject matter is the improvement of human welfare. Nevertheless, their critics have often called them a heartless crew, content with the calculus of more and less within the existing order, while so much of humanity suffers and dies, and the gross sins of society go unstudied and uncorrected.

Economists' public esteem, to be sure, has had its ups and downs. In the first two decades following World War II—a period of rapid and steady growth, high employment, and improving welfare in the United States and most other countries—the prestige of economists soared to an all-time high. But the economists' once exalted reputation for worldly wisdom has been tarnished of late. They generally (Boulding and Galbraith were exceptions) failed to anticipate the deterioration of the environment stemming from rising production, population growth, and technological advance. Nor did they foresee the impact of rapid economic change on America's cities.

The economists were preoccupied with economic growth and full employment, but the worrisome lesson of the past decade is that growth and full employment are no answers to what ails the United States and other advanced industrial societies. The rise in affluence, the growing strength of economic organizations, the unwillingness of societies to suffer mass unemployment once the Keynesian remedy was at hand

—all this has made inflation a chronic problem in the West. The rise of affluence is fine for those who have benefited from it, but its side-effects have been the intensification of social discontents and antisocial behavior within the developed countries, and a growing antagonism between the rich nations and poor nations.

Only yesterday, Nazism and Fascism, feeding on economic insecurity, led to the most horrible war in history. Rising tensions in Africa, the Middle East, the Caribbean, and even in Western Europe are fresh reminders that the preservation of political and economic freedom—and of peace itself—depends significantly on economic security and well-being. For us, over the long run, the conflicts between rich and poor nations loom as a chronic danger. People will not accept "freedom" as a substitute for food, clothing, shelter, and a measure of security. Indeed, it is a cruel paradox that political repression should be used to protect "freedom," because working people will not tolerate unemployment and rampant inflation: this is the lesson of Chile after 1973. In the wake of the overthrow of Marxist President Salvador Allende Gossens, the Chilean junta sought to cure its economic ills by applying the free-enterprise and monetarist solutions of Milton Friedman. But in following these policies, the junta stifled the news media, banned political parties, and shackled labor unions.

Capitalism and freedom are not exactly inseparable. Economic institutions never exist in a political vacuum; every economy is imbedded in a particular polity. The illusion that economics could be divorced from politics existed for a while in England and the United States during the nineteenth and early twentieth centuries—a spell of unusually fair weather for laissez-faire capitalism that has now blown away.

This is a time of much soul-searching among economists over the philosophical and methodological bases of their discipline. Many economists today would still maintain that economics is simply a set of analytical techniques, applicable to certain aspects of production, distribution, exchange, and consumption. They would assert that the science of economics aims at increasing the efficient use of resources and realizing the goals postulated by individuals, businesses, or governments, but that economists as such cannot specify the goals for society. Their science, they insist, gives them no means and no special qualifications for doing so, and it would be elitist or authoritarian for them to try.

That economics can or should be value-free is an antiseptic view which is not shared by Boulding, Galbraith, Leontief, Samuelson, or implicitly even by Friedman, all of whom have strong, though different, value positions which they are willing to assert in behalf of society.

A still stronger insistence upon a specific set of values, especially those of social justice and equality in the distribution of income, is coming today from the "radical economists," who are often disciples of Marx but who commonly oppose authoritarian political rule. They reject what they consider to be the complacent acceptance of the existing American political-economic system by so-called establishment or mainstream economists.

Of our five economists, only Samuelson is in the very center of the mainstream. Friedman is to the right; Galbraith and Leontief have shifted to the left and are fairly close to the radical or new left economists. Boulding, with his religious belief in love and the perfectibility of man, is distant from the radicals and their dialectics (he rejects conflict as a creative force). But Boulding, though a lover of freedom, is a long way from Friedman's intense concentration on the market as the source of all virtue or from Samuelson's agnosticism, willingness to live within the boundaries of traditional economics, and lack of interdisciplinary zeal or hope.

The radical economists contend that the professionally cool and neutral establishment economists have, wittingly or unwittingly, helped to support institutions that serve the well-to-do and the private corporations, not the broad public weal. The radicals charge that mainstream economics is the economics of the bourgeoisie. Galbraith, close to the radicals on this point, charges that establishment economists are not worldly but bourgeois philosophers held in thrall, and helping to hold the common people in thrall, by their elaborate system of belief which masquerades as economic science.

Keynesianism, which once seemed so revolutionary as doctrine, is seen by the radicals as simply a means of propping up capitalism. It does not reach the power structure that shapes society in the interests of the powerful, those who control the great corporations and, through them, the government and the people.

But it is not only the radicals who sense that the age of Keynes is over. Sir John Hicks, a Nobel laureate in economics, has suggested that the third quarter of the twentieth century was indeed the age of Keynes, an epoch of prolonged boom that ended in stagnation and rapid inflation.[1] Hicks called the second quarter the age of Hitler, the epoch of fascism, war, and genocide. The century's first quarter, which Hicks has not titled, was perhaps the age of Henry Ford, the era of unbridled industrial capitalism, mass production and mass consumption.

Obviously, each quarter is not one of a kind, a product of one man who put his mark on it. Rather, each of the quarters appears to be a counterstroke to the quarter that preceded it. Thus, the second quarter's age of totalitarianism and war was the consequence of the breakdown of the old industrial order and of the mass unemployment that laissez-faire economics could not cure. The third quarter's age of Keynes was, in turn, a counterstroke to the disasters of the second, an effort of the capitalist democracies to prevent a recurrence of mass unemployment without resort to either fascism or war. For at least half of the third quarter, the Keynesian revolution was a great success; through the 1950s and most of the 1960s, the industrial world experienced what economists call a "long wave" of expansion, during which recessions were shallow and brief.

It is an open question how much of the long postwar boom can be attributed to Keynesian policies. Some countries went Keynesian, others did not. But as Professor Hicks contends, the impact of the nations that adopted Keynesian policies was apparently strong enough to have caused a general boom. Of course, there were other factors to explain the boom, especially the combination of more rapid technological progress and the increased "socialist" demand of nations for collective goods (not only social welfare programs but also military adventures in Korea and Vietnam, major weapons programs, and such quasi-military developments as nuclear energy and the space program).

Nevertheless, in the minds of many, the long postwar boom was linked with the new Keynesian doctrine. This was the cause of the rising prestige of the new economics, which reached a peak at the end of the Kennedy administration and the start of the Johnson administration in the United States. By the same token, however, with the faltering of the boom and the soaring of inflation, the Keynesian policies have come under heavy attack. Many economists contend that the Keynesian doctrines were essentially designed to deal with the problems of the Depression, when deep unemployment was combined with deflation rather than (as now) inflation. They consider that the basic Keynesian tool of "demand management" is no longer satisfactory for dealing with stagflation—that is, stagnation and inflation combined.

Stagflation seems to call for greater focus on the supply side, to increase the availability of resources and the factors of production. However, this does not mean abandoning efforts to control total demand. What economists are looking for is a way to operate on both sides of the aggregate supply-demand equation. Economic policy needs to focus on structural economic problems, such as programs to improve labor skills, to increase investment in research and development, to expand the production of raw materials, to safeguard the environment, to enhance living conditions in urban areas, to reduce the amount of idle manpower (including those on welfare), and to provide better incentives for productive economic activity.

"The purely Keynesian era is over," says Neil J. McMullen of the National Planning Association.[2] "What is needed may not amount to a revolution in economics but is certainly an evolution in economic thinking—that is, the development of macroeconomic policy beyond demand management toward the augmentation of supplies and the enhancement of incentives." And Robert J. Gordon of Northwestern University observes: "A good argument can be made that the day of stabilization policy is past and that the real progess in the next ten to twenty years will be in the area of dealing with the allocation of resources and the distribution of wealth and income."[3]

There has been widespread antipathy in recent years among politicians and the general public toward the economists, the great majority of whom continued to focus on the manipulation of demand. This rising antipathy has been by no means confined to the United States. In Britain, the cradle of classical and neoclassical economics running from Adam Smith through Lord Keynes, the *New Statesman* in 1976 awarded a prize for the following definition of "economist":

An inhabitant of cloud-cuckoo land; one knowledgeable in an obsolete art; a harmless academic drudge whose theories and laws are but mere puffs of air in face of that anarchy of banditry, greed and corruption which holds sway in the pecuniary affairs of the real world.[4]

To redeem their reputation throughout the industrial
world in the years ahead, economists must move on beyond
the economics of Smith and Keynes. Yet many of the leaders
of the economics profession are reluctant to move.

I I

THE LEADERS of the economic establishment still consider
economics to be the queen of the social sciences, and by far
the most successful of them. They contend that economics
has attained this eminence by sticking to problems which it
can handle.

Professor James Tobin of Yale says that the agenda for
economics derives from two sources: (1) exogenous sources,
that is, the troubles of the real world, such as unemployment
or inflation; and (2) endogenous sources, or the internal
momentum of the science of economics itself.[5] If economists
pay too much attention to the endlessly changing outside
problems, they run the risk of being faddish; if they concentrate too much on their internal intellectual problems, they
run the risk of growing increasingly esoteric and irrelevant to
society. Tobin suggests that a "golden age" occurs for economics whenever there is a convergence of the external and
internal agenda. Such a convergence occurred at the end of
the eighteenth century and the beginning of the nineteenth,
when "external" political and business controversies over international trade, economic development, and the protection-
ist corn laws in Britain coincided with "internal" development of the theory of markets by the economists. Another
golden age for economics came in the 1930s, when the Great
Depression came together with the theories of Keynes on how
the economy as a whole functions and with the statistical

work on national income by Simon Kuznets at the National Bureau of Economic Research.

Does a new golden age for economics lie just ahead? The economists are dubious. There is a lot of excitement among the economic theorists—especially the young ones—about abstract matters. But these seem more and more remote from the real-world problems that overlap economics and other disciplines: the population explosion, the apparent shift of power to the oil-producing countries, the growing conflict between the poor nations and the rich, world industrialization and hunger, threats to the environment and the exhaustion of nonrenewable resources, and the dangers of war and nuclear holocaust. Traditional economics is ill-equipped to deal with such problems.

This divergence between the external and internal agenda poses a dilemma for economists: "Whether to have more and more to say about less and less, or . . . less and less to say about more and more," as Professor Robert Solow of MIT puts it. For his part, Professor Solow chooses the former alternative, and suggests that it would be better for economists to aspire to be competent technicians, "like dentists," as Keynes once put it. But this is a choice that a number of dissident economists now regard as a fate worse than death, a retreat from the world's really important problems. Galbraith, as we have seen, concedes that economists "can, if they are determined, be unimportant; they can, if they prefer a comfortable home life and regular hours, continue to make a living out of the infinitely interesting gadgetry of disguise. . . ."[6] They can answer questions that anyone but mathematicians or quiz-show contestants would avoid. But if that is all they do, they will be "socially more irrelevant than Keynes's dentist, for he would feel obliged to have a recommendation, were everyone's teeth, in conflict with all expectation, suddenly to fall out." The alternative is for economists to enlarge their system, to have it embrace "the power they now disguise."[7]

E Pluribus Unum

Many other economists, and not only the radicals, have called for broadening the scope as well as the methods of economics, lest economists look too narrowly at problems and get absurd results as a consequence of the way their subject matter is defined. Professor Daniel Fusfeld of the University of Michigan has pointed, for instance, to the economists' great concern over the recycling of Arab oil profits, to the complete neglect of what he considers the real story: "How one elite rips off another, and how the second elite defends itself."[8] Professor Robert Heilbroner of the New School for Social Research calls for fresh thinking by economists on such issues as population growth, controls to prevent an environmental catastrophe, corporate power (especially that of the multinationals), the spread of nuclear capabilities, and the possibility that the world has entered a period of decline like that of the Roman Empire, with everything starting to come apart—including long-standing value systems.[9]

The more traditional economists find all such talk vague, and beyond the scope of economics. They insist that the present use of economic tools, though limited, is at least "productive," and they urge the economics profession not to throw away its hard-won methodology.

Samuelson's famous (some thought it infamous) remark on this subject was delivered as the last line of his presidential address to the American Economic Association: "In the long run, the economic scholar works for the only coin worth having—our own applause." He later explained that he meant this to be "a plea for calling shots as they really appear to be . . . even when this means losing popularity with the great audience of men and running against 'the spirit of the times.' "[10]

Professor Solow says that people delight in comparing economists to the drunk who lost his key and kept looking for it under the street light, because at least he could see there. This, he suggests, is a joke on the people who tell the story; he feels economists *should* keep looking where they have some light. But others, like Professor Leontief, are fed up with what they consider time-wasting studies of more and more about less and less.

Clearly, the establishment that dominates the American economics profession is not eager to change its ways, either in response to outside criticism or to pressure from dissidents within the profession, and it is rearing up a new generation in its own image. Professor Tobin insists that his own economics department knows how to produce "plumbers"—that is, competent economic technicians—but it does not know how to produce "philosopher-kings." He is doubtful, in any case, about the substantive value of the philosopher-kings: "Any honest man or woman owes it to the reader to say why you should believe what I say," says Tobin. "Can it be proved empirically? Can it be observed?"[11]

Specifically, Tobin argues that those economists who like to talk about "power" should produce an operationally useful definition of power; any economics journal would be glad to publish such an article. It has not been done, he suggests, because it is too difficult—and he himself has tried.

But is it really that hard? A working definition of power is the ability of anybody to make somebody else do something that he does not want to do. Could this be measured in the case of an oil cartel which levies higher prices on oil-consuming countries? In the case of a country which nationalizes the property of a multinational corporation? In the case of a second country which seeks to prevent the first from nationalizing the property of its multinationals?

Or are these problems that economists do not like to handle because they do not fit neatly into the existing body of neoclassical economic theory and analysis? Or because, as Galbraith charges, they would disturb the political peace in which most economists like to live?

The left contends, of course, that power resides in the eco-

nomic class that dominates the society and controls the distribution of income and wealth, not the marginal productivity
of labor, capital, and other factors of production, as traditional economic theory maintains. That is the central point of
Karl Marx. It was not exactly a new discovery, even when
Marx made it over a century ago. In his famous revolutionary sermon at Blackheath, at the time of the Peasant Revolt
of 1381, the lay preacher John Ball declared: "They have
wine and spices and fair bread; and we have oat-cake and
straw, and water to drink. They have leisure and fine houses;
we have pain and labor, the rain and the wind in the fields.
And yet it is of us and of our toil that these men hold their
state."[12]

But one does not have to be a peasant revolutionary or a
Marxist to recognize that class—and race and national origin
—have a good deal to do with the distribution of income, at
any given time and over long periods of time. Conventional
economists, through most of the past century, have concentrated too narrowly on increasing the efficiency and growth
of production, while paying too little attention to the distribution of wealth and income, or to the still wider issue of
social justice. They have assumed that efficiency was the obvious (or implicit) objective of their discipline, but that the
study of social and economic justice was beyond the reach of
"economic science." The economists, with rare exceptions,
treated equality and inequality, justice and injustice, as matters for philosophers, lawyers, or the citizenry in general, but
not for themselves qua economists.

But such recent and contrasting works as John Rawls's *A
Theory of Justice* and Robert Nozick's *Anarchy, State, and
Utopia*—together with the heightened turmoil and conflict in
the world over equal opportunity and income distribution—
have aroused economists to the need of addressing themselves
to the critical questions of justice and equality, and how these
questions relate to such other economic objectives as effi-

ciency, growth, welfare, and freedom. If economists would presume to minister to the ills of society, as they do, they cannot simply select those goals or values that conveniently fit the traditional concepts and models and data of their discipline.

III

ECONOMICS has often been called the "science of choice." Classical economics is orderly, definite, linear, clear—like classical music. Here, says the professor of conventional economics, are the factors of production: land, labor, and capital. Decision makers combine these elements to maximize something, whether it be efficiency, production, short-term profits, long-term profits, or growth. But what happens when research, technological change, innovation, new ideas, *new knowledge* (emphasized particularly by Professor Boulding, Galbraith, and Leontief, as we have seen) become major factors in increasing production and productivity, thereby changing the nature of civilized life? The focus then shifts from the neat strings of factors of production to complex and unique human personalities and—in a corporate, governmental, or academic world where the creative process is to some extent institutionalized—to large-scale human organizations.

Hence, argue the critics of conventional economics, the walls of economic theory must be broken down to let in the other social sciences. Instead of making the simplifying assumption that man is a pleasure/pain, profit/loss calculating machine, economists must look more profoundly at the question of what really makes human beings behave as they do.

The need for a broader and deeper mode of analysis in a

number of economic and social realms is beginning to be recognized. Economists once thought, for instance, that to bring a backward economy into the modern age, the essential job was to transfer capital to it and to generate capital internally. This, many now see, is a laughably inadequate answer. The range of social, psychological, educational, entrepreneurial, political, and other complexities involved in the economic development of poor countries defy the simplistic assumptions of conventional economists. And the persistence of poverty and the worsening of many social and environmental problems in rich, highly developed societies should force economists to question the adequacy of their tools for improving human welfare—the classic aim of economics.

In my view, economics unquestionably needs a more realistic conception of human welfare than is provided by data on income and output per capita. "What constitutes the well-being of a man?" Carlyle asked in 1839. In part, wages and the amount of bread his wages will buy. But these, he said, were only the preliminaries:

> Can the labourer, by thrift and industry, hope to rise to mastership; or is such hope cut off from him? How is he related to his employer—by bonds of friendliness and mutual help, or by hostility, opposition, and chains of mutual necessity alone? In a word, what degree of contentment can a human creature be supposed to enjoy in this position?
>
> With hunger preying on him, his contentment is likely to be small! But even with abundance, his discontent, his real misery may be great. The labourer's feelings, his notion of being justly dealt with or unjustly; his wholesome composure, frugality, prosperity in the one case, his acrid unrest, recklessness, gin-drinking, and gradual ruin in the other—how shall figures of arithmetic represent all this?[13]

How indeed? Intangible but important goods such as justice and friendship are not touched upon by the arid arithmetic of economics. But economists are beginning to recognize that such factors are as crucial to real human well-being as those goods more readily fitted with price tags.

In an effort to devise and set forth measurements of such goods, some scholars have sought to develop social indicators to set beside the economic indicators as measures of welfare and guides to social policy. The overall aim is a numerical scale or set of scales that will give a truer picture of welfare than, say, the gross national product alone. Paul Samuelson, drawing on the work of William Nordhaus and James Tobin, has produced a revised version of the GNP called the net economic welfare (NET), which, for example, takes into account the product of household work and corrects for some of the disamenities of urbanization, such as pollution and environmental costs.[14]

Quantifying human welfare in this manner is incredibly difficult, because welfare is a complex cultural bundle, not a series of discrete phenomena. Is welfare necessarily increased if people have longer vacations but the highways are choked, the beaches overcrowded, and the costs of summer houses near the beach incredibly high? Does "more" produce greater welfare, or does "less"? More is not necessarily better, but less may not be better either, if it means less health, less freedom of movement in the city, less music, less friendship, less work at jobs people really want to do.

There lies another great complication: welfare is subjective, and people want very different things. Many want big, fast cars, electric appliances and gadgets, the pleasures of conspicuous consumption, the slaying of wild animals, the speed of moving in supersonic planes. Others want the quiet life and freedom from all the above. Is it possible to reconcile the existing range of subjective preferences with the welfare of all? Does the freedom to choose ensure greater subjective welfare, even for the individual, let alone for society? What if one chooses heroin? What degree of freedom is optimal for mankind? Jefferson offers one view, Dostoevsky another,

Sartre a third, Camus a fourth, Leonid Brezhnev a fifth, Mao Tse-tung a sixth.

Samuelson, who has devoted much of his career to trying to get rid of the rocks on the unpromising terrain of welfare economics, has shown how the interaction of people seeking to realize their subjective preferences increases the welfare of society as a whole, yielding a "social welfare function" that can be optimized, at least theoretically. But Galbraith has warned us against the power of those who control the production processes of large organizations to manipulate others in such a way as to *reduce* their welfare; he has attacked what he regards as the myth of consumer sovereignty, contending that it is the big producers who are sovereign. Boulding sees love and peace as the most important aspects of human welfare.

The falsely simple method of adding and subtracting things in order to measure "what constitutes the well-being of a man," as Carlyle put it, actually seems to be giving way to a richer and truer understanding among economists than the simplistic utilitarianism that long characterized economics. After many years of smugness about the state of their art, more and more economists are asking themselves broader questions about what they should be doing, and how. This may ultimately have a great deal to do with how people and nations can be helped to behave more decently and sensibly, with less harm to one another.

Perhaps some danger exists that economics—a limited field, but one with some significant achievements to its credit —will drown prematurely in a sea of related disciplines before it has adequately solved some of its more traditional problems, such as how to achieve both full employment and price stability, or how to put together a stable, essentially free, expanding world economy. Yet it seems to me that efforts to solve even those traditional economic problems cannot be hampered but only advanced by greater understanding of many matters that lie beyond the boundaries of conventional economics.

IV

TO SOLVE the highly complex problems that have come to the fore in modern societies, economists need to escape from the straitjacket imposed upon economic thought by an uncritical use of mathematical concepts and symbols. Percy Bridgman, a distinguished mathematician, philosopher, and physicist, has observed that while there is no sharp distinction between "mathematics" and "verbalizing," our traditional verbal habits may nevertheless have "the highest guiding and constructive value." The concepts actually used in physics (or economics) and the operations which give them meaning are, as Bridgman says, only a few of the enormous number of conceivable concepts and operations:

> It is no accident that so many times we are able, by giving heed merely to our verbal demands, to evolve a concept or point of view that is relevant to an "external" physical situation. For our verbal habits have evolved from millions of years of searching for adequate methods that were not a close enough fit. The desirability of continuing to use our old verbal habits in new situations if possible is obvious enough in achieving economy of mental effort, and the probability of at least a partial success is suggested by our universal experience that absolutely sharp breaks never occur, but that a method, hitherto inadequate, can always be extrapolated beyond its present range with some partial validity. *By the same token, however, the validity of any extrapolation may be expected eventually to break down, so that one may anticipate ineptnesses or inadequacies in concepts which have been formed by too uncritical a verbal extrapolation.*[15] [Italics added.]

Bridgman suggests that it would be a good idea to re-examine all the concepts of physics from a verbal point of view to discover how each concept evolved, whether its utility has been circumscribed by its origin, and to what extent it is misused because of its "verbal bar sinister." I feel confident that this applies with at least equal force to economics, since, as John von Neumann and Oskar Morgenstern said in their *Theory of Games*, "our knowledge of the relevant facts of economic life is incomparably smaller than that commanded by physics at the time when the mathematization of that subject was achieved." What, for example, is *capital*? Joan Robinson opened up a highly significant debate by showing that a particular equation produces highly different results depending on how capital is defined. In such cases, the "precision" of mathematical language is illusory. As the philosopher Wittgenstein said, ordinary language is frequently a better guide to the understanding of new and complex problems than are mathematical symbols, precisely because ordinary words are somewhat vague, general, and capable of shifting their meaning and of embracing in a subtle way many of the more difficult issues that philosophy—or economics—confronts.[16] Wittgenstein held that the illusion that for each word (or symbol) there exists a crystal-clear, sharply defined meaning creates the further illusion that, because everyday language employs words and sentences that lack this crystal-clear language, it must be inadequate or somehow insufficient for the uses of serious thinkers.

But abstract economic theory is heady stuff for those who delight in its beauty—and, apart from its aesthetic value, it may sometimes even be useful. It is virtually impossible to know in advance whether any piece of abstract work will or will not prove useful for solving practical problems. The history of mathematics includes both kinds of cases, as does economics. For example, linear programming (to which Samuelson has made important contributions) has proved extremely valuable in a wide range of practical applications, such as where an oil company should locate its refineries or what is the best mix of products for a manufacturer to produce. Some of the advanced theoretical work being done in economics today may have value for future philosophers or scientists, even though it has little relevance to the problems of inflation, unemployment, or economic development.

It is the nature of basic research that one does not know whether, when, or where the payoffs will come. This is particularly true for the economist, who is at a disadvantage relative to the natural scientist in the firmness with which he can establish generalizations or "laws." The natural scientist frequently can test his theories rigorously in a laboratory under controlled conditions, but the laboratory of the economist is, with rare exceptions, the real world with all its complexities and experimental "contaminants." However, in this respect, the work of the economist differs from that of the natural scientist only in degree, not in kind. What is more important is that the economist and the natural scientist share in the creative process of discovering the secrets of nature or society. Indeed, what these scientists do is closely akin to what the *artist* does. The artist searches out hidden likenesses and creates order out of chaos. In the words of the late Jacob Bronowski, philosopher of both science and literature:

> We re-make nature by the act of discovery, in the poem or in the theorem. And the great poem and the deep theorem are new to every reader, and yet are his own experiences, because he himself re-creates them. They are the marks of unity in variety; and in the instant when the mind seizes this for itself, in art or in science, the heart misses a beat.[17]

V

ECONOMICS may be closer to art than to science.

Nicholas Georgescu-Roegen of Vanderbilt University, a distinguished mathematician and economist, has noted the important differences between the models of economics and of the natural sciences. In physics, he says, the model is a *calculating device*, from which the scientist can compute the answer to any question regarding the physical behavior of the corresponding physical system; such a model represents an *accurate blueprint* of a particular sector of physical reality. But an economic model, as Georgescu-Roegen demonstrates, is not an accurate blueprint but an *analytical simile*.[18]

The difference between a *blueprint* and a *simile*, as Georgescu-Roegen sees it, is that anyone can follow a blueprint (as in assembling a radio apparatus purchased in kit form) and automatically achieve the desired results; but no one, not even a consummate economist, can depend on an economic model as a guide to *automatic action*. A simile states a likeness between two essentially unlike things: here is a host of people buzzing around a grain pit, buying and selling different lots of grain at different prices, and here are an intersecting supply curve and demand curve which exhibit a certain likeness to what is happening in the grain pit. But to use the supply-and-demand model of economics for practical purposes, says Georgescu-Roegen, requires "delicacy and sensitivity of touch"—"call it art, if you wish." An artless analysis, he adds, cannot subserve an art.[19]

Thus, in the physical sciences the model must be accurate in relation to the sharpest measuring instrument available at the time, whereas in the social sciences there is no comparable objective standard of accuracy, "no acid test for the validity of an economic model." This is why economists often get such wildly disparate results essentially from the same data.

Barbara Bergmann of the University of Maryland offered a devastating example of this in her presidential address at the inaugural meetings of the Eastern Economic Association:

> Developments in economics that look extremely promising frequently turn out to be dross. In 1971, for example, the Brookings Institution published a book edited by Gary Fromm with the title *Tax Incentives and Capital Spending* which reported on a conference of experts which had been held in 1967. The experts who came together included Robert Hall of M.I.T. and Dale Jorgenson of Harvard who produced a paper which featured the theory of optimal capital accumulation, Charles R. Bischoff of Yale who produced a paper emphasizing lags in investment response, Robert M. Coen of Stanford who produced a paper featuring the influence of cash flows on investment behavior, and Lawrence R. Klein and Paul Taubman of the University of Pennsylvania who produced a paper in which the question at issue was studied in the context of a complete econometric model.[20]

All the papers, Professor Bergmann noted, represented the "very best product that the economics profession has to offer." But there was one problem, which Franklin M. Fisher of M.I.T., a participant in the conference, put this way:

> The four analyses presented in this book are all marked by high quality. Each applies sophisticated econometric tools to the empirical and theoretical analysis of an important problem; each does so in a professional and convincing manner; each sheds light where before there was darkness. If it were not for the inconvenient fact that the four analyses happen to concern the same problem and happen to contradict each other's findings, there would be little to discuss. Except for that, indeed, the contribution of each to economic science and to public policy seems assured.[21]

This is no isolated or trivial case. Bergmann thinks the answer is better data—more direct first-hand observation of reality and more data generation by economists, who are virtually the only would-be scientists who are content to use stale, second-hand data collected chiefly by government agencies for other purposes. Economists are even willing to ignore the gross inaccuracy of much of their data, as Oskar Morgenstern has made brilliantly clear in *On the Accuracy of Economic Observations.*

The best economists are aware of the importance of trying to observe life directly, to see what is really there, in any way they can. The young Leontief in China sends out an airplane to photograph the fields and rice paddies; the mature Leontief uses his eyes and ears and nose in observing the garbage dumps of Tokyo, the cane fields of Cuba, the factories and schools and railroads of China, and he listens to people. Leontief in turn praises Marx for his "direct observation of the capitalist system" in nineteenth-century Britain. Samuelson notes that Arthur Burns must have been sensitive to "the cigarette smoke" in automobile salesrooms in 1954; cigarette smoke may tell you more—and sooner—than the published data can. Galbraith travels, rolls up his pants, gets his feet wet; conducting a strategic bombing survey, he finds that the bombing had far less effect than advertised in destroying a highly developed industrial system in Germany. If only the U.S. Air Force and the White House had remembered and believed the study—and realized how much more difficult it would be to destroy the resistance of the more backward, rural economy of Vietnam through aerial bombing tactics!

Barbara Bergmann would send monetary economists out to talk with bankers; she says a well-known Princeton monetary economist who claimed he had never learned a thing from bankers was incapable of listening to what the bankers said. There is a whole world of people waiting to be observed and *listened* to. Economists who choose to enter that world may be astonished at what they learn—and at how much of existing economic theory is thereby shattered.

What the economist always needs are fresh and apt models, which require genuine artistry to create, and the skill to use existing models imaginatively and insightfully.

Economics becomes artistry when Boulding uses the model of a bathtub, with resources pouring in and not draining out, as a means of shedding light on a fundamental cause of war, which comes when society must "pull the plug" so that economic accumulation can start again. The simile may not be far-fetched; World War II was followed by a long period of rebuilding capital equipment that brought with it rapid growth and high prosperity, but that period may now have reached an end.

V I

IN MY VIEW, the libertarians are seeking to extend the domain of the market beyond what is appropriate to an economically healthy and socially just society. Libertarians like Friedman are trying to revive the system of social values and economic institutions that characterized the United States prior to the New Deal, before the United States and other capitalist nations set out on what Friedrich von Hayek called "the road to serfdom."

Economic freedom (which the libertarians take as the necessary condition for all freedom) and social justice are really quite different values. Economic freedom is the right of individuals or businesses to act autonomously, free of government constraint or direction (with highly limited exceptions). Social justice can most simply be defined as the principle that equals will be treated equally—and, just as important, the corollary that unequals will be treated unequally. (Thus, for example, the principle of equity in taxation holds that all people with the same income should pay the same taxes; and people with higher or lower income should pay higher or lower taxes. Or, under just law, people who are found guilty of committing the same crime should receive equal punishment, but people who commit a greater offense should receive greater punishment.)

The crucial question in deciding what constitutes social justice is determining the essential features of equality or inequality. Obviously, people are not equal in all respects; some are taller, heavier, more intelligent, or more energetic than others. From the standpoint of law, we in the United States maintain that all citizens are equal and should receive equal treatment (except for minors or certain types of mentally retarded people, who are treated differently). But while we say that all people are entitled to equal justice, there has been no presumption in our society that all are entitled to equal economic rewards. We assert that rewards should be proportional to economic contribution. Yet many of our institutions (such as the progressive income tax, free public education, welfare payments, food stamps) imply a national belief that the distribution of income should be made *somewhat* more equal than market forces alone would yield, and we put a floor under the income of many individuals to prevent starvation and utter destitution.

Is this as far as we should go? Or can we go further in redistributing the fruits of our highly productive economic system without undermining its freedom and efficiency? Here is one of the critical, emergent issues of our time, and one that takes many forms, embracing not only such economic issues as tax reform, inheritance laws, welfare payments, and minimum wages but also such social issues as busing, school integration, and "benign quotas" for minority groups in education and employment.

True-blue conservatives have little trouble with such issues. Anything that impairs the efficiency or freedom of the system, they say, is wrong—economically, socially, and morally. They are in the tradition of that nineteenth-century champion of laissez-faire, Herbert Spencer, who declared, "The command, 'if any would not work neither should he eat,' is simply a Christian enunciation of that universal law of nature under which life has reached its present height—the law that a creature not energetic enough to maintain itself must die." Spencer was even against free public libraries, on the ground that they gave people something for nothing and encouraged loafing. Latter-day Spencerians see the pursuit of equality as sapping the life of capitalist growth, with socialist Britain the most horrible example of economic decay.

On the other side, egalitarians see ever-expanding market capitalism as the corrupter of human dignity and human rights. Money can buy things that should not be for sale, including justice and political power, and thereby vitiate our formal national commitments to equal justice under the law and to the equal democratic rights of citizens. The market, notes Arthur Okun of Brookings, is even permitted "to legislate life and death, as evidenced, for example, by infant mortality rates for the poor that are more than one and one-half times those for middle-income Americans."[22]

Yet there is a good case for the market, in terms of both efficiency and freedom. Private ownership circumscribes and constrains the power of government. As Paul Samuelson has remarked, nothing so increased his own enthusiasm for this virtue of private enterprise than did the "McCarthyism" of the early 1950s. In the 1970s, liberals' wariness of the powers of the state was reawakened by the administration of Richard Nixon. This was not, however, a simple matter of state power versus the private sector, for a close alliance existed between many "private enterprisers" and the government, with favors and money passing back and forth. A corrupt symbiosis of private wealth and political power would, if it were to continue, obviously threaten the life of democracy itself.

The market, if it can be kept honest and competitive, provides strong incentives for work effort and productive contributions. Wassily Leontief, through advocating socially oriented economic planning, seeks to preserve such self-interested incentives. He offers as his model for society this simile: "sails" of profit catch the wind and move the ship of state, while the "rudder" of planning gives the economy and society better direction.[23] He is wary of altruism or collective loyalty as unreliable incentives—and fearful that coercion or oppression will replace them when they fail.

But one can't always have the best of everything, and the problem facing society today may be whether to sacrifice a certain amount of efficiency for greater equality. Arthur Okun scoffs at those conservative intellectuals who, he says, forget the Declaration of Independence, ignore the Bill of Rights, and investigate egalitarianism "as though it were an idiosyncrasy, perhaps even a type of neurosis." He offers the simile of a "leaky bucket" by which society might transfer income from the rich to the poor, though a great deal of the income might simply leak out and be lost before it ever got to the poor.

Should the transfer of income by leaky bucket be made anyway? John Rawls would give a clear, crisp answer: make the transfer to increase equality, unless an unequal distribution of income would definitely be to everyone's advantage; otherwise, equality should have the priority. Milton Friedman would give an equally clear and crisp but opposite answer: give the priority to efficiency. Okun's own answer isn't neat; he would fall somewhere between the two positions. He says capitalism and democracy need each other, "to put some rationality into equality and some humanity into efficiency."[24]

But it is the American society itself, not the economists, that will have to decide how to solve this leaky-bucket problem, or how to reconcile the domain of rights and the domain of dollars. Societal judgments on the way to reconcile these claims, though oscillating, have shifted somewhat toward equality as total national income has grown.

The shift was anticipated and partly brought about by Galbraith, who argued so brilliantly that however appropriate a very limited degree of social redress of economic equality might have been in an earlier age of general poverty and scarcity, such extremes of inequality were no longer appropriate, necessary, or just in an affluent society. The lesson has still not sunk in deeply enough.

Nor can we any longer justify our continued neglect (or devastation) of the social and natural environment in the name of market efficiency or growth. Fortunately, equality and decency are not always in conflict with efficiency. Herbert Spencer was almost certainly wrong, for instance, about the real effect of public libraries; they have surely done more to improve "human capital" than to breed sloth. And the output lost from the waste of human resources is almost certainly greater than the social costs needed to develop them. Efficiency, in any case, is not an ultimate but only an instrumental value. It is only a mindless and compulsive society that could make efficiency the be-all and end-all of economic and social action. Economics appears to be shifting away from such mindlessness, and a good thing it is.

VII

DESPITE their professional failings and their slanging matches over political and economic issues—and even despite the widespread public criticism and abuse to which they are periodically subjected—the economists remain prosperous and apparently influential.

Why is this so? Barbara Bergmann suggests that economists' advice is "something like patent medicine—people know it is largely manufactured by quacks and that a good percentage of the time it won't work, but they continue to buy the brand whose flavor they like."[25]

Whether the flavor of economic advice you like is conservative or liberal, you will find that flavor available from some "reputable" economist since there is no single standard to which all "reputable" economists must repair. The lack of this single standard means that conservative economists are free to give advice which, if taken, would make things better at least in the short run for business and for the rich. Economists who are left of center are free to give advice which, if taken, will in the short run tend to help the lower income groups. Thus right-wing groups can always find economists to say (on their oath as "scientists") that the way to fight inflation is to lower the tax on business profits, while the left-wing will always find economists to say on an equally scientific basis that tax breaks for the lower income groups is the correct policy in an inflation. Each group has thus the possibility of pretending that economic science endorses that group's own self-interested demands. . . .[26]

Professor Bergman cruelly suggests that economists' bread and butter is thus assured for a long time to come, despite or perhaps because of the fact that they are not increasing the body of useful knowledge on which all can agree. But she concedes that at least many economists have aspirations which go higher than "selling themselves to the interest groups or doing exercises in applied mathematics of which the applications never actually occur."

It really seems to be impossible to disentangle economists' social philosophies or politics from their economics. Was this what the Swedish Academy of Science was saying when it awarded the 1974 Nobel Prize in Economic Science simultaneously to Gunnar Myrdal, a socialist, and Friedrich von Hayek, a libertarian? In fact, both Myrdal and Hayek have stressed the difficulty, or impossibility, of purging economic analysis of political, social, and moral values. Myrdal has contended that "problems in the social sciences—not only the practical ones about what ought to be done, but also the theoretical problems of ascertaining the facts and the relations among facts—cannot be rationally posited except in terms of definite, concretized and explicit value premises."[27] And Hayek, in his Nobel lecture, attacked what he called the "scientistic" attitude of economists, an approach which he said is "decidedly unscientific in the true sense of the word, since it involves a mechanical and uncritical application of habits of thought to fields different from those in which they have been formed."[28]

The public is usually not deceived by some economists' pretension to scientific purity; it identifies economists as far right, conservative, middle-of-the-road, liberal, or left-wing, and knows that this often (though not necessarily) affects the nature of their analyses or recommendations. But the economist is sometimes self-deceived, and only succeeds in lowering his own standing and that of his profession by seeking to invoke professional authority for his personal values.[29]

There is, after all, nothing wrong with controversy among economists—indeed, it is a crucial part of the process by which truth is discovered and values explicated.

This many-sided discourse among the economists has never been more intense than it is today. Paul Samuelson notes that the "new economics"—that is, the economics of Keynes and Samuelson that remains the mainstream of the economics profession—is now under attack on four fronts. First there are the conservative interests, whose historic rejection of efforts to use government economic policy to promote high employment and price stability, says Samuelson, "lacks intellectual interest—either you feel it or you don't." Second come the libertarians, such as Milton Friedman, who have reminded us of what it is that market pricing accomplishes, and what are some of the penalties to society of disregarding these lessons." Third is the unique assault by John Kenneth Galbraith, to whose crusade for greater attention to programs in the public sector Samuelson himself says "Right on!" Fourth and finally comes the new left, "the children of affluence, [who] even after the Vietnam war is blissfully behind us, will, according to the timetable forecast by Schumpeter thirty years ago, turn increasingly critical of the established system."[30] Children of affluence the new left may be, but they are also the children of Marx, taunting the economics establishment for overlooking or underrating the roles of class and power in the economic system. They stand outside the gates of the establishment, but their voices are heard. And so, too, are the voices of moral and ethical concern, such as those of Kenneth Boulding and Wassily Leontief.

Out of these many voices does one economics emerge? In a strange way, it does: a dialogue of knowledgeable combatants who know each other's tricks; a contest among intellectuals for the public influence they sometimes possess.

At their best and most disinterested, the economists have sought to understand a mysterious and vital phenomenon—the economy—and to prescribe for its real and serious ills; to shape events rationally in an effort to advance society's welfare, as they have variously interpreted it; and to teach society and statesmen the truth. These are high and ambitious goals which, however short of them economists may have fallen, make far more sense than trusting to the blind forces of history.

NOTES

Paul Anthony Samuelson

1. Thorstein Veblen, "The Intellectual Pre-eminence of Jews in Modern Europe," *Political Science Quarterly* 34 (March 1919): 33–42; reprinted in *Essays in Our Changing Order,* ed. Leon Ardzrooni (New York, 1934); and in Leonard Silk, *Veblen: A Play in Three Acts* (New York, 1966), p. 8.

2. Paul A. Samuelson, "Economics in a Golden Age: A Personal Memoir," in *The Twentieth-Century Sciences: Studies in the Biography of Ideas,* ed. Gerald Holton (New York, 1972), p. 161.

3. Samuelson, "A Note on Measurement of Utility," *Review of Economic Studies* 4 (February 1937): 155–161. This essay and all other scientific essays published by Professor Samuelson between 1937 and 1971 have been reprinted in *The Collected Scientific Papers of Paul A. Samuelson,* vols. 1–2, ed. Joseph E. Stiglitz (Cambridge, Mass., 1966); vol. 3, ed. Robert C. Merton (Cambridge, Mass., 1972).

4. Samuelson, "Some Aspects of the Pure Theory of Capital," *Quarterly Journal of Economics* 51 (May 1937): 469–496.

5. Samuelson, "A Note on the Pure Theory of Consumer's Behavior," *Economica* 5 (February 1938): 61–71.

6. Samuelson, "Welfare Economics and International Trade," *American Economic Review* 28 (June 1938): 261–266.

7. Samuelson, "The Empirical Implications of Utility Analysis," *Econometrica* 6 (October 1938): 344–356.

8. Samuelson, "Interactions between the Multiplier Analysis and the Principle of Acceleration," *Review of Economics and Statistics* 21 (May 1939): 75–78; © 1939 by the President and Fellows of Harvard College.

9. Samuelson, *Foundations of Economic Analysis* (Cambridge, Mass., 1948).

10. Samuelson, "Foreword" to the Japanese edition, *Foundations of Economic Analysis* (1967).

11. Samuelson, "Economics in a Golden Age," p. 170.

12. Samuelson, "Maximum Principles in Analytical Economics" (Nobel Memorial Prize Lecture, 1970), *Science* 173 (September 1971): 996–997.

13. Samuelson, *Economics: An Introductory Analysis,* 1st ed. (New York, 1948). Subsequent editions have been published every three years, through the 10th ed. (1976).

14. Samuelson, "Intertemporal Price Equilibrium: A Prologue to the Theory of Speculation," *Weltwirtschaftliches Archiv* 79 (December 1957): 181–219.

15. Ibid., 209.

Notes

16. Samuelson "Personal Freedoms and Economic Freedoms in a Mixed Economy," in *The Business Establishment*, ed. E. F. Cheit (New York, 1964), pp. 624–626.

17. Wassily Leontief, "Introduction" to *The Research Revolution*, by Leonard Silk (New York, 1960), pp. 6–8.

18. Samuelson, "Economics in a Golden Age," pp. 160–161.

19. Samuelson, "Comment on Ernest Nagel's 'Assumptions in Economic Theory,'" *Papers and Proceedings of the American Economic Association*, December 29, 1962, p. 1774.

20. Ibid.

21. Thomas Kuhn, *The Structure of Scientific Revolutions* (Chicago, 1962, 1970).

22. This was the "correct" answer to a *New York Times* history test question, published May 2, 1976, pp. 1, 65.

23. Samuelson, "Modern Economic Realities and Individualism," in "Individualism in Twentieth-Century America," ed. Gordon Mills, *Texas Quarterly* (Summer 1963): 137.

24. Milton Friedman, *Capitalism and Freedom* (Chicago, 1962).

25. Samuelson, "Modern Economic Realities and Individualism," 135–136.

26. Lewis Thomas, *The Lives of a Cell* (New York, 1974).

27. Samuelson, "Maximum Principles in Analytical Economics," 997.

Milton Friedman

1. Milton Friedman, "The Economic Theorist," originally published under the title "Wesley C. Mitchell as an Economic Theorist," *Journal of Political Economy* (December 1950); reprinted in *Wesley Clair Mitchell, the Economic Scientist*, ed. Arthur F. Burns (New York, 1952), pp. 237–238.

2. Paul A. Samuelson, "Harold Hotelling as Mathematical Economist," *American Statistician* 14 (June 1960): 21.; reprinted in *The Collected Scientific Papers of Paul A. Samuelson*, (Cambridge, Mass., 1966), vol. 2, p. 1588.

3. Friedman, *Consumer Expenditures in the United States* (Washington, D.C., 1939), p. 830.

4. Friedman, *Studies in Income and Wealth* (New York, 1938), vol. 2, p. 123.

5. Ibid., p. 127.

6. Ibid., pp. 129–130.

7. Friedman, "Preface" to *Capitalism and Freedom* (Chicago, 1962).

8. Friedman and Simon Kuznets, *Income from Independent Professional Practice* (New York, 1945), p. 11.

9. Ibid., p. 137.

10. Friedman et al., *Taxing to Prevent Inflation* (New York, 1943), pp. 83–84.

11. Ibid., p. 131.

12. Ibid., pp. 136–137.

Notes

13. Ibid., p. 140.

14. Friedman, *Essays in Positive Economics* (Chicago, 1953), p. 253.

15. Ibid., p. 253, fn. 2.

16. Friedman, *An Economist's Protest* (Glen Ridge, N.J., 1972), pp. 15–16.

17. Friedman, "The Spendings Tax as a Wartime Fiscal Measure," *American Economic Review* 33 (March 1943): 54.

18. Friedman, *An Economist's Protest*, p. 158.

19. Ibid., p. 15.

20. Paul H. Douglas, *In the Fullness of Time* (New York, 1972), pp. 127–128.

21. Friedman and George Stigler, *Roofs or Ceilings? The Current Housing Problem* (Great Barrington, Mass., 1946), p. 9.

22. Ibid., p. 10.

23. Ibid., p. 10.

24. Friedman, *Essays in Positive Economics*.

25. Ibid., p. 4.

26. Ibid., p. 6.

27. Ibid., pp. 14–15.

28. Ibid., p. 39.

29. Ibid., p. 39.

30. Ibid., p. 39.

31. Ibid., pp. 41–42.

32. Friedman's use of simple models, based on past experience, is well illustrated by the OPEC case. In June 1974 he wrote: "The tendency for the power of a cartel to decline is reinforced by a . . . key economic proposition: the more successful the cartel, the greater the pressure on it not only from a reduction in consumption and an increase in noncartel supply but also from within the cartel. The members of a cartel have conflicting interests. Each is anxious for the others to cut production while he expands production to profit from the high prices. The process has been very evident in the oil cartel. Libya, Iran, Iraq, and some other members of OPEC have been vociferous in demanding ever higher prices while themselves producing all out to benefit from the high prices. They have been able to succeed so far because they have been able to persuade Saudi Arabia and Kuwait to bear the whole of the cut in production. But the cuts required by Saudi Arabia and Kuwait will become larger and larger, and their willingness to make them smaller and smaller, as the downward pressure on prices from cuts in consumption and alternative supplies increases. The cartel will break up—as just about every other such cartel in the past has—long before Saudi Arabia and Kuwait output approach zero. . . . These considerations explain why, in my opinion, the 'energy crisis' was greatly exaggerated, why it will disappear from our major concerns and why the price of oil will return to a level much closer to its pre-October 1973 price than to the peak prices reached shortly thereafter" (Milton Friedman, *There's No Such Thing as a Free Lunch* [Chicago, 1975], pp. 307–308). However, other cartel members—including not only Saudi Arabia but even such a poor country as Nigeria—have been willing to cut production to hold the cartel price high and to keep the cartel together. The revival of the world economy after the 1973–1975 slump would revive the demand for oil and strengthen the

Notes

cartel's bargaining power. Friedman was willing to make his confident prediction of the drop in the oil price back close to the pre-October 1973 level and the breakup of the oil cartel without concerning himself with complex studies of world demand for oil or the impending exhaustion of petroleum reserves in the coming half-century. He could also ignore the political cement among the oil producers: the anti-Israeli purposes of the Arab states and the effort among cartel members as a whole to make common cause with the other developing countries against the rich industrial states. A simple market model presumably made such considerations superfluous.

33. Nicholas Georgescu-Roegen, *Analytical Economics* (Cambridge, Mass., 1966), p. 117.

34. Friedman and Anna J. Schwartz, *A Monetary History of the United States, 1867–1960* (Princeton, 1963).

35. Friedman, *Essays in Positive Economics*, pp. 41–42.

36. See Don Patinkin, "The Chicago Tradition, the Quantity Theory, and Friedman," *Journal of Money, Credit and Banking* 1 (February 1969): 46–70.

37. Abba P. Lerner, "Keynesianism—an Exaggerated Demise and a Premature Funeral," Conference on the Relevance of the New Deal to the Present Situation, City University of New York, June 23–25, 1975.

38. Friedman, *Capitalism and Freedom*, p. 54.

39. Friedman, "Some Comments on the Significance of Labor Unions for Economic Policy," in *The Impact of the Union* (New York, 1951).

40. *New York Times*, October 28, 1971.

41. Friedman, "Have Monetary Policies Failed?" American Economic Association convention, New Orleans, December 28, 1971.

42. Friedman, *Capitalism and Freedom*, p. 169.

43. Ibid., p. 21.

John Kenneth Galbraith

1. *[MORE]* 4 (October 1974): 20.

2. John Kenneth Galbraith, "Berkeley in the Thirties," in *Economics, Peace and Laughter* (Boston, 1971), p. 347.

3. Leonard S. Silk, "The Problem of Communication," Papers and Proceedings of the 66th Annual Meeting of the American Economic Association, Boston, December 27–29, 1963, in *American Economic Review* 54 (May 1964): 595–609.

4. E. C. Hughes, "Professions," *Daedalus* (Fall 1963): 657.

5. Galbraith, "Berkeley in the Thirties," p. 349.

6. Ibid., pp. 351–352.

7. Galbraith, "Monopoly Power and Price Rigidities," *Quarterly Journal of Economics* 50. (May 1936): 456–475.

8. Galbraith and John D. Black, "The Quantitative Position of Marketing in the United States," *Quarterly Journal of Economics* 49, (May 1935): 394–413.

9. Galbraith, "Monopoly Power," 468.

Notes

10. Ibid., p. 473.
11. Ibid., p. 475.
12. Ibid., p. 473.
13. Henry S. Dennison and Galbraith, *Modern Competition and Business Policy* (New York, 1938).
14. Morris E. Leeds, Ralph E. Flanders, Lincoln Filene, and Henry S. Dennison, *Towards Full Employment* (New York, 1938).
15. Galbraith, "How Keynes Came to America," in *Economics, Peace and Laughter*, p. 50.
16. Galbraith, *A Theory of Price Control* (Cambridge, 1952), p. 17.
17. *New York Times*, June 6, 1943, sec. 4, p. 8.
18. *New York Times*, May 28, 1943, p. 29.
19. *New York Times*, May 25, 1943, p. 1.
20. See Galbraith, "A Retrospect on Albert Speer," in *Economics, Peace and Laughter*, pp. 288–302.
21. The two earlier articles were "Reflections on Price Control," *Quarterly Journal of Economics* 60 (August 1946), 475–489, and "The Disequilibrium System" *American Economic Review* 37 (June 1947): 287–302.
22. F. W. Taussig, "Price-Fixing as Seen by a Price Fixer," *Quarterly Journal of Economics* 33 (February 1919): 205–241.
23. Arnold Weber, "Making Wage Controls Work," *The Public Interest* (Winter 1973): 28–40.
24. Galbraith, *Theory of Price Control*, p. 4.
25. Ibid., p. 7.
26. Ibid., p. 9.
27. Ibid., p. 24.
28. Ibid., p. 60.
29. Ibid., p. 71.
30. V. S. Navasky, "Galbraith on Galbraith," *New York Times Book Review*, June 25, 1967, p. 3.
31. Plato, *Gorgias*, 454D–455A.
32. Galbraith, *American Capitalism*, 2nd ed. (Boston, 1962), p. 9.
33. Ibid., p. 17.
34. Ibid., p. 55.
35. Ibid., p. 96.
36. Ibid., p. 111.
37. Ibid., p. 170.
38. Ibid., p. 200.
39. Ibid., p. 201.
40. Plato, *Gorgias*, 459C.
41. Galbraith, *The Great Crash* (Boston, 1954).
42. Galbraith, *Economics and the Art of Controversy* (New Brunswick, N.J., 1955).
43. Ibid., 2nd ed. (New York, 1959), p. ix.
44. Ibid.
45. Galbraith, *The Affluent Society*, paperback ed. (New York, 1958), p. 18.
46. Ibid., p. 22.
47. Ibid., p. 116.
48. Ibid., p. 124.

Notes

49. Ibid., p. 196.
50. Ibid., p. 267.
51. Ibid., p. 271.
52. Galbraith's diverse output includes: *Journey to Poland and Yugoslavia* (1958), a journal; *The Liberal Hour* (1960), a collection of essays; *The McLandress Dimension* (1962, 1963), his pseudonymous satire; *The Scotch* (1964), reminiscence; *How to Get Out of Vietnam* (1967), a political pamphlet; *Indian Painting* (1968), a history written with M. S. Randhawa; *The Triumph* (1968), a novel; *Ambassador's Journal* (1969), account of the Indian years; *How to Control the Military* (1969), a political pamphlet; *Economics, Peace and Laughter* (1971), a collection of essays; *A China Passage* (1973), a journal.
53. Galbraith, *Ambassador's Journal*, paperback ed. (New York, 1970), p. xiv.
54. Galbraith, *The New Industrial State* (Boston, 1967), p. ix.
55. Ibid., p. 7.
56. Ibid., p. 12.
57. Ibid., pp. 23–24.
58. Thornstein Veblen, *The Engineers and the Price System* (1921; reprint ed., New York, 1963), pp. 64–67 and passim.
59. Galbraith, *New Industrial State*, pp. 58–59.
60. Ibid., pp. 168–169.
61. Ibid., p. 376.
62. Ibid., p. 61.
63. Ibid., p. 399.
64. Galbraith, *Economics and the Public Purpose* (Boston, 1973), p. ix.
65. Ibid., p. 324.
66. Galbraith, "Power and the Useful Economist," reprinted in Leonard Silk, *Capitalism: The Moving Target* (New York, 1974), pp. 136–156.

Wassily Leontief

1. Unless otherwise indicated, direct quotations from Professor Leontief are drawn from tape-recorded interviews with him by the author and by Robert Silvers and Emma Rothschild, who have kindly made their transcripts available to me.
2. First published in *Weltwirtschaftliches Archiv* 22 (October 1925): 338–344, and *Weltwirtschaftliches Archiv Chronik und Archivalien* 22 (1925): 265–269. Russian translation, "Balans narodnogo khoziaistva SSSR—metodologicheskii razbor rabotii TSSU," *Planovoe Khoziaistvo*, no. 12 (Moscow, 1925): 254–258. English translation in N. Spulber (ed.), *Foundations of Soviet Strategy for Economic Growth: Selected Short Soviet Essays, 1924–1930* (Bloomington, 1964), pp. 88–94.
3. William Henry Spiegel, *The Growth of Economic Thought* (Englewood Cliffs, N.J., 1971), p. 200.

Notes

4. Wassily Leontief, "Quantitative Input and Output Relations in the Economic System of the United States," *Review of Economic Statistics* 18 (August 1936): 39–59.

5. Leontief, *The Structure of the American Economy, 1919–1929* (Cambridge, Mass., 1941).

6. Leontief, "Input-Output Economics," *Scientific American* 185 (October 1951): 8.

7. W. Isard and Phyllis Kaniss, "The 1973 Nobel Prize for Economic Science," *Science* 182 (November 9, 1973): 569–571.

8. Leontief, "Sails and Rudders, Ship of State," *New York Times*, March 16, 1973; reprinted in *Capitalism: The Moving Target*, comp. Leonard Silk (New York, 1974), pp. 101–104.

9. Leontief, "Primer for the Great Society," *New York Review of Books*, December 15, 1966.

10. Leontief, "The Structure of Development," *Scientific American* 209 (September 1963): 148–164.

11. Leontief, "The Rates of Long-Run Economic Growth and Capital Transfer from Developed to Underdeveloped Areas," in *Essays in Economics: Theories and Theorizing*, (New York, 1966), pp. 200–215.

12. Leontief, "Theoretical Note on Time-Preference, Productivity of Capital, Stagnation and Economic Growth," in ibid., pp. 175–184.

13. Leontief, "An Institute for Technical Economics," *Bulletin of the Atomic Scientists* 21 (September 1965): 46.

14. Leontief, "Environmental Repercussions and the Economic Structure: An Input-Output Approach," *Review of Economics and Statistics* 52 (August 1970): 262–271.

15. Leontief, "Mysterious Japan: A Diary," *New York Review of Books,* June 4, 1970, pp. 23–29.

16. Leontief, *Essays in Economics*, p. viii.

17. W. Leontief, "Theoretical Assumptions and Non-Observed Facts," *American Economic Review* 61 (March 1971): 1.

18. Robert Dorfman, "Wassily Leontief's Contribution to Economics," *Swedish Journal of Economics* 75 (1973): 430–449.

19. Leontief, "Bigger and Better?" *New York Review of Books*, May 4, 1972, page 35.

20. Leontief, "Introduction" to *The Research Revolution*, by Leonard Silk (New York, 1960), pp. 1–8.

21. W. Leontief, "Bigger and Better?" 34–35.

22. Leontief, "The Significance of Marxian Economics for Present-Day Economic Theory," *Essays in Economics*, p. 78.

23. Leontief, "The Decline and Rise of Soviet Economic Science," *Essays in Economics*, p. 224.

24. Leontief, "The Limits of Economics," *New York Review of Books*, July 20, 1972, p. 32.

25. Leontief, "Decline and Rise of Soviet Economic Science," 225.

26. Ibid., 226.

27. Leontief, "Notes on a Visit to Cuba," *New York Review of Books*, August 21, 1969, p. 16.

28. Leontief, "The Trouble with Cuban Socialism" (Morristown, N.J., 1973), pp. 1–2; reprinted from *New York Review of Books*, 1971.

29. Ibid., p. 6.

Notes

30. Leontief, "Socialism in China," *Atlantic Monthly* (March 1973): 74.

31. Ibid., pp. 80–81.

32. Ibid., p. 81.

Kenneth E. Boulding

1. Kenneth E. Boulding, *There Is a Spirit* (Sonnet no. 11, in Naylor Sonnets) (New York, 1945), p. 11.

2. Howard Brinton, *Friends for 300 Years* (New York, 1952).

3. Boulding, *Collected Papers*, vol. 1 (Boulder, Colo., 1971), pp. viii–ix.

4. Boulding, "The Place of the 'Displacement Cost' Concept in Economic Theory," *Economic Journal* 42 (March 1932): 137–141; and Boulding, *Collected Papers*, vol. 1, p. 6.

5. Ibid., p. 4.

6. Cynthia Earl Kerman, *Creative Tension: The Life and Thought of Kenneth Boulding* (Ann Arbor, 1973), p. 121.

7. Boulding, *Economic Analysis* (New York, 1941), p. xviii.

8. Quoted in Kerman, *Creative Tension*, p. 121.

9. Boulding, Sonnet no. 1, in *There Is a Spirit* (Naylor Sonnets), p. 3.

10. Pierre Teilhard de Chardin, "The Mysticism of Science" (1939), reprinted in *Human Energy*, trans. J. M. Cohen (New York, 1969), p. 164.

11. Ibid., p. 180.

12. Boulding, "New Nations for Old," Pendleton Hill Pamphlet no. 17 (Wallingford, Pa., 1942), pp. 20–28.

13. Boulding, "The Practice of the Love of God," William Penn Lecture (Philadelphia, 1942), pp. 20–21.

14. Boulding, "Collective Bargaining and Fiscal Policy," *American Economic Review* 40 (May 1950): 306–320; also *Collected Papers*, vol. 1, pp. 275–291.

15. Boulding, *Collected Papers*, vol. 1, p. xi.

16. Boulding, *The Economics of Peace* (New York, 1945), p. 252.

17. Ibid., p. v.

18. Ibid., p. 181.

19. Boulding, "The Consumption Concept in Economic Theory," *American Economic Review* 35 (May 1945): 1–14; also in *Collected Papers*, vol. 1, pp. 153–168.

20. Boulding, *A Reconstruction of Economics* (New York, 1950), p. 194.

21. Ibid., p. vii.

22. Ibid., pp. vii–viii.

23. Derek de Solla Price, *Science since Babylon* (New Haven, 1961), pp. 54, 99.

24. See Pierre Teilhard de Chardin, "The Formation of the Noo-

Notes

sphere" (1947), reprinted in *The Future of Man*, trans. Norman Denny (New York, 1964), pp. 161–191.

25. Boulding, "A Conceptual Framework for Social Science," *Papers of the Michigan Academy of Science, Arts and Letters* 37 (1951); also in Boulding, *Beyond Economics* (Ann Arbor, 1970), pp. 57–63, and in *Collected Papers*, vol. 4 (forthcoming).

26. Boulding, "Economics as a Social Science," in *The Social Sciences at Mid-Century: Essays in Honor of Guy Stanton Ford* (Minneapolis, 1952), pp. 70–83; also in *Collected Papers*, vol. 3 (forthcoming).

27. Boulding, "Toward a General Theory of Growth," *Canadian Journal of Economics and Political Science* 19 (August 1953): 326–40; reprinted in Boulding, *Beyond Economics*, pp. 64–82, and in *Collected Papers*, vol. 3.

28. Boulding, "Contributions of Economics to the Theory of Conflict," *Bulletin of the Research Exchange on the Prevention of War*, 3 (May 1955): 51–59; also in *Collected Papers*, vol. 5.

29. Boulding, "The Malthusian Model as a General System," *Social and Economic Studies* 4 (September 1955): 195–205; also in *Collected Papers*, vol. 1, pp. 451–463.

30. Boulding, "Notes on the Information Concept," *Exploration* 6 (1955): 103–112; also in *Collected Papers*, vol. 4.

31. Boulding, "Protestantism's Lost Economic Gospel," *Christian Century* 67 (1950): 970–972.

32. Boulding, "Religious Perspectives of College Teaching in Economics," in *Religious Perspectives of College Teaching*, ed. Hoxie N. Fairchild (New York, 1952), pp. 360–383; also in Boulding, *Beyond Economics*, pp. 179–197.

33. Boulding, "Religious Foundations of Economic Progress," *Harvard Business Review* 30 (May–June 1952): 33–40; reprinted in *Beyond Economics*, pp. 198–211, and in *Collected Papers*, vol. 3.

34. Boulding, "The Quaker Approach in Economic Life," in *The Quaker Approach*, ed. John Kavanaugh (New York, 1953), pp. 43–58.

35. Boulding, *The Organizational Revolution* (1953; reprinted., Chicago, 1968), p. xvii.

36. Ibid., p. xxxv.

37. Ibid., p. 49.

38. Ibid., pp. 67–68.

39. Ibid., p. 178.

40. Ibid., p. 198.

41. Boulding, "The Practice of the Love of God," p. 3.

42. Boulding, *Organizational Revolution*, p. 218.

43. Ibid., p. 217.

44. Ibid., p. 193.

45. Ibid., pp. 220–221.

46. Boulding, "Organization and Conflict," *Journal of Conflict Resolution* 1 (June 1957): 122–134; reprinted in *American National Security: A Reader in Theory and Policy*, ed. Morton Berkowitz and P. G. Bock (New York, 1965), pp. 327–336.

47. Boulding, "National Images and the International System," *Journal of Conflict Resolution*, 3 (June 1959): 120–131; also in *Collected Papers*, vol. 5.

Notes

48. Boulding, *The Image* (Ann Arbor, 1956), p. 18.

49. Ibid., p. 155.

50. Boulding, "A New Look at Institutionalism," in *Collected Papers*, vol. 2, p. 100; this article originally appeared in the *American Economic Review* 47 (May 1957): 1–12.

51. Boulding, "Economic Libertarianism," in *Conference on Savings and Residential Financing, 1965: Proceedings* (Chicago, 1965), pp. 30–42; also in *Beyond Economics*, pp. 43–54.

52. Boulding, "Notes on a Theory of Philanthropy," in *Philanthropy and Public Policy*, ed. Frank G. Dickinson (National Bureau of Economic Research, 1962), pp. 57–71.

53. Boulding, "The Grants Economy," *Papers of the Michigan Academy of Science, Arts and Letters*, 1 (Winter 1969): 3–11; also in *Collected Papers*, vol. 2, pp. 475–485.

54. Boulding, *The Economy of Love and Fear; A Preface to Grants Economics* (Belmont, Cal., 1973), p. 47.

55. Boulding, *The Skills of the Economist* (Cleveland, 1958), pp. 182–183. These lectures were originally given in Brazil in August and September 1953.

56. Boulding, *Conflict and Defense* (New York, 1963), p. 248.

57. Boulding, *Economics as a Science* (New York, 1970).

58. Boulding, "Some Contributions of Economics to Theology and Religion," in *Beyond Economics*, p. 223. This article originally appeared in *Religious Education* (December 1957): 446–450; also in *Collected Papers*, vol. 4.

59. Boulding, *The Meaning of the Twentieth Century* (New York, 1964), p. 192.

60. Boulding, *A Primer on Social Dynamics* (New York, 1970), p. 62.

61. Mubashir Hasan, address before the annual meeting of the International Monetary Fund and World Bank, Washington, D.C., October 1, 1974.

62. Leonard Silk, *The Research Revolution* (New York, 1960), p. 50.

63. M. Hasan.

64. *Everyman and Medieval Miracle Plays*, ed. A. C. Cawley (New York, 1959), p. 13.

E Pluribus Unum

1. Sir John Hicks, *The Crisis in Keynesian Economics* (New York, 1974), pp. 1–3.

2. Quoted in Leonard Silk, "Some Find Keynes Policies Outmoded," *New York Times*, April 21, 1976, pp. 49, 53.

3. Ibid., p. 53.

4. *New Statesman*, March 12, 1976, p. 339.

5. Quoted in Leonard Silk, "Paradox for Economists," *New York Times*, October 30, 1974, p. 63.

Notes

6. John Kenneth Galbraith, *Economics and the Public Purpose* (Boston, 1973), p. 324.

7. Ibid., p. 324.

8. Quoted in Silk, "Paradox for Economists," p. 63.

9. Robert L. Heilbroner, *Business Civilization in Decline* (New York, 1975).

10. Paul A. Samuelson "Economists and the History of Ideas," Presidential Address to the American Economic Association, in *American Economic Review* 52 (March 1962):18.

11. Quoted in Silk, "Paradox for Economists," p. 63.

12. J. R. Green, *A Short History of the English People* (New York, 1916), p. 250.

13. Thomas Carlyle, "Chartism," in *English and Other Critical Essays* (New York, 1915), pp. 172–173.

14. Paul A. Samuelson, *Economics*, 9th ed. (New York, 1973), pp. 195–197.

15. P. W. Bridgman, *The Nature of Thermodynamics* (New York, 1961), pp. x–xi.

16. Arne Naess, *Four Modern Philosophers* (Chicago, 1968), pp. 149–153.

17. J. Bronowski, *Science and Human Values* (New York, 1965), p. 20.

18. Nicolas Georgescu-Roegen, *Analytical Economics* (Cambridge, Mass., 1966), p. 116.

19. Ibid., pp. 116–117.

20. Barbara R. Bergmann, "Have Economists Failed?" (Presidential Address to the Eastern Economic Association, Albany, N.Y., October 27, 1974, pp. 4–6.

21. Ibid., p. 5.

22. Arthur M. Okun, *Equality and Efficiency* (Washington, D.C., 1975), p. 22.

23. Wassily Leontief, "Sails and Rudders, Ship of State," in *Capitalism: The Moving Target*, ed. Leonard Silk, pp. 101–104.

24. Okun, *Equality and Efficiency*, p. 120.

25. Bergmann, "Have Economists Failed?" p. 3.

26. Ibid., pp. 3–4.

27. Gunnar Myrdal, *Challenge to Affluence* (New York, 1963), p. vi; see also his *Value in Social Theory* (New York, 1958).

28. Friedrich A. von Hayek, "The Pretense of Knowledge," Nobel Memorial Lecture, Stockholm, December 11, 1974; see also his "Scientism and the Study of Society," *Economica*, 9 (August 1942): 267–291.

29. Leonard Silk, "The Problem of Communication," Papers and Proceedings of the 66th Annual Meeting of the American Economic Association, Boston, December 27–29, 1963, in *American Economic Review* 54 (May 1964): 595–609.

30. Samuelson, "Liberalism at Bay," Second Gerhard Colm Memorial Lecture, New School for Social Research, March 5, 1971.

INDEX

Index

Index

Economics as a Science (Boulding), 230

Economics of Imperfect Competition, The (Robinson), 101–102

Economics of Peace, The (Boulding), 208–210

Economists: public esteem of, 243; radical, see Radicals

Economy of Love and Fear, The (Boulding), 228–229

Education: Galbraith on, 131, 132, 140–141; Leontief on, 179–180

Educational voucher system, Friedman on, 84–85

Egalitarianism: Samuelson and, 35–36; see also Equality, economic

Eiconics, Boulding on, 225–226

Engineers and the Price System, The (Veblen), 138

Environmental problems, Leontief on, 160, 172–173

Equality, economic, 265–268; Boulding on, 217; Friedman on, 70; see also Inequality, economic

Equation of exchange, 62, 202

Essays in Positive Economics(Friedman), 65, 71–72, 77

Ethical norms or values (moral values), 269, 270; Boulding on, 191–192, 216, 218, 221; Friedman on, 57, 71–72; Samuelson on, 35–40

Evans, W. Duane, 163

Evolutionary process of history, Boulding on, 231–234

Exchange, equation of, 62, 202

Exchange rates, Friedman's advocacy of floating, 84

Family Assistance Program, 85

Farm prices, 106; Galbraith on, 101–103

Federal Reserve System, 81, 82, 87

Filene, Lincoln, 106

Fisher, Franklin M., 262

Fisher, Irving, 62, 202

Flanders, Ralph E., 106

Ford, Gerald, 170–171

Forecasting, economic: Friedman on validity of hypotheses and, 72–74; Leontief's input-output analysis and, 164, 165; by monetary models, of short-run changes, 64; Samuelson's approach to, 33–35

Fortune (magazine), 111

Foundation for Economic Education, 69–70

Foundations of Economic Analysis (Samuelson), 3, 12, 13

France, Anatole, 37

Free market, Friedman's views on, 61, 69–70, 73, 74

Freedom, 244; economic, 264; Friedman's view of capitalism and, 92–93

Friedman, Milton, 5, 47–93, 166–167, 244, 245, 267; Boulding and, 227–228; on capitalism's achievement, 92–93; at Chicago University, 50–53; at Columbia University, 53–54, 60, 67; on deficit budgets, 63–64, 77–78; early life of, 48–50; on economic theory, 49–50, 74–75; on equality, economic, 70; on ethical norms, 57, 71–72; on forecasting, 72–74; on free market, 61, 69–70, 73, 74; on freedom, 92–93; on full-employment budget, 64, 77–78; Goldwater candidacy and (1964), 87–88; on income tax, 77, 85, 86; on inflation, monetary factors in, 62–65, 78–81; on inflation, Nixon policy on, 89–91; on inflation, taxation and, 61–63, 65–67; on inflationary gap, 62–65; Keynesian economics and, 65, 76–80; on labor unions, 82–83; on laissez-faire, 47–48, 57, 79, 80, 82, 86; Mitchell and, 49–50, 53–55, 58, 76; monetarist theory of, 76, 78–83; on monetary policy, 77, 78, 90–91; on money supply, growth of, 76, 80–81, 83; on money supply, inflation and, 63–64; on monopolistic competition, 74; on monopoly, 82; National Bureau of Economic Research and, 55,

Index

Index

Index

Index

Military expenditures: Galbraith on, 116, 117, 130; Leontief on, 173

Mill, John Stuart, 37, 201

Mises, Ludwig von, 86

Mitchell, Wesley C., 49–50, 55, 58, 76, 157

Mixed economy, Samuelson and, 24–25, 41

Models, mathematical, Leontief on, 175–176

Modern Competition and Business Policy (Galbraith and Dennison), 105

Modern Corporation and Private Property, The (Berle and Means), 102

Monetarism, Friedman's theory of, 76, 78–83

Monetary History of the United States, A (Friedman), 76–77, 87

Monetary policy, Friedman on, 77, 78, 90–91

Money, quantity theory of, 62, 78

Money supply: Friedman on growth of, 76, 80–81, 83; Friedman on inflation and, 63–64; *see also* Money, quantity theory of

Monopolistic competition (or imperfect competition), 28, 31, 102; Friedman on, 74; Galbraith on, 102–104; Samuelson and theory of, 28–31

Monopoly: Friedman on, 82; Galbraith on, 102–105, 120–122, 124; of knowledge, Samuelson on, 27–28; *see also* Antitrust legislation

Moore, Marianne, 189

Moral incentives, Leontief on, 171, 183

Moral issues, *see* Ethical norms or values; Normative economics

Morgenstern, Oskar, 230, 259, 263

Morton, Walter, 59

Myrdal, Gunnar, 39–40, 269–270

Mysticism, Boulding and, 205–206

Nachman, Gerald, 97

National Bureau of Economic Research, 49, 55, 59–61, 76, 157–158

National Resources Committee, 54

Nature and Significance of Economics, The (Robbins), 35

Negative income tax, Friedman on, 85

Net economic welfare (NET), 256

New Class, Galbraith on, 131, 141

New Economic Policy, 89

New Industrial State, The (Galbraith), 136–141

New left, 270–271; *see also* Radicals

New socialism, Galbraith's, 143–145

News Statesman, 248

Newsweek (magazine), 88–89

Niebuhr, Reinhold, 222

Nixon, Richard M., 20–21, 67, 77–78, 85, 88–89, 180

Normative economics: Samuelson and, 35–36; *see also* Ethical norms or values

Nozick, Robert, 253

Office of Price Administration (OPA), Galbraith at, 107–111

Okun, Arthur, 266, 267

Oligopoly, Galbraith on, 102–104, 121

Organizational Revolution, The (Boulding), 216–222

Organizations, Boulding on, 216–222, 224–225

Output: real, Friedman on technological change and, 56; *see also* Input-output analysis; Production

Oxford University, Boulding at, 194–197

Pacifism, Boulding and, 200, 204–208, 223–224, 236

Paish, F. W., 63

Paradigms, economic, Samuelson on, 30–31

[291]

Index

Index

Religion, Boulding on, 192,
230–231, 234
Rent control, Friedman on, 69
Republican Party, 87
Ricardo, David, 197, 201
Richardson, Lewis, 223, 229–230
Richardson, Stephen, 223
Risk, Uncertainty and Profit
(Knight), 51
Robbins, Lionel, 35, 195
Robinson, Joan, 28, 52, 65, 74,
101–102, 259
Rogin, Leo, 100
Romer, Alfred, 9
Roofs or Ceilings? (Friedman and
Stigler), 69–71
Roosevelt, Franklin D., 110

Salant, Walter, 65
Salesmanship, Galbraith on, 121,
129, 137–138
Samuelson, Marion, 12, 33
Samuelson, Paul A., 3–43, 201, 245,
251, 266, 270–271; early life and
background of, 4–8; on economic
paradigms, 30–31; on ethical
norms or values, 35–40; on
forecasting, 33–35; Friedman
and, 19, 29–30, 37–38, 52, 71; as
graduate student, 10–13; on
information, uses of, 23–25;
Keynesian economics and, 12,
16–17; on knowledge, 25–28;
laissez-faire doctrine and, 24,
27–30; mathematics and, 41–42;
at MIT, 14, 16; mixed economy
and, 24–25, 41; monopolistic
competition theory and, 28–31;
on monopoly of knowledge,
27–28; normative economics and,
35–36; on perfect competition,
28–31; positivistic economics
and, 29–30, 35; radicalism and,
18–20; social-welfare goal and,
28, 35–36, 256, 257; on
speculation, 23–25; summary
evaluation of, 41–43; as
undergraduate, 8–10; on
unemployment, 17, 39
Samuelson, Robert, 7

Scarcity, Galbraith on, 127, 128,
130
Schultz, Henry, 51
Schultz, Theodore, 18, 208
Schuman, Frederick, 9
Schumpeter, Joseph, 14–15, 26,
121, 198
Schwartz, Anna J., 87
Science: economics as, Friedman
on, 71–72 (*see also* Positivistic
economics, Friedman on);
mysticism and, Teilhard de
Chardin on, 205–206; natural,
relationship between economics
and, 260–261; *see also* Social
sciences
Security, economic, Galbraith on,
127–129, 137, 140
Self-interest: Friedman on, 47;
Leontief on, 166–167, 170, 171,
183, 267; Smith on, 28, 37, 47;
see also Profit motive
Shoesmith, Beulah, 7
Shoup, Carl, 61–62, 66
Shultz, George, 89
Simons, Henry, 51, 69, 77
Smith, Adam, 28, 36–37, 201;
Friedman and, 47, 48
Social balance, Galbraith on,
130–131
Social indicators, 256
Social sciences, Boulding on
integration of, 212, 213, 215, 226,
227
Social welfare (human welfare):
need for new conception of,
255–257; Samuelson on, 28,
35–36, 256, 257
Socialism: Boulding and, 197;
Galbraith on, 143–145
Socialists, 127
Society for General Systems
Research, 222
Socrates, 118, 205
Solow, Robert, 250–252
Soviet economists, 4, 182
Soviet Union, Leontief on, 182,
185–187
Speculation, Samuelson on, 23–25
Speer, Albert, 112

Index